Explain
that

Also available from *The Age* and
The Sydney Morning Herald Explainer desk

What's It Like to Be Chased by a Cassowary? (ed. Felicity Lewis)

Explain that

31
intriguing reasons
WHY

Edited by
Felicity Lewis

Illustrated by Dionne Gain

PENGUIN BOOKS

UK | USA | Canada | Ireland | Australia
India | New Zealand | South Africa | China

Penguin Books is part of the Penguin Random House group of companies whose addresses can be found at
global.penguinrandomhouse.com

Penguin
Random House
Australia

First published by Penguin Books in 2021

Chapter openers and cover illustrations by Dionne Gain
Cover design by James Rendall © Penguin Random House Australia Pty Ltd
Internal design by James Rendall © Penguin Random House Australia Pty Ltd
Typeset in 11/16 pt Abril Text by Post Pre-press Group, Brisbane

Printed and bound in Australia by Griffin Press, part of Ovato, an accredited
ISO AS/NZS 14001 Environmental Management Systems printer

 A catalogue record for this
book is available from the
NATIONAL
LIBRARY National Library of Australia
OF AUSTRALIA

ISBN 978 1 76104 446 5

penguin.com.au

Contents

INTRODUCTION

Felicity Lewis

If you could travel in time, which way would you go? In these pages, we've become time tourists, returning to the beech forests of Gondwana to discover how our quirky plants and animals evolved (giant wombats were real), chasing peas around our plates at dinner with a sixteenth-century French king to skewer the origins of table manners (there's a reason you don't put your elbows on the table, kids) and wandering through nineteenth-century Sydney, listening for the early twangs of a distinctly Australian accent.

But is actual time travel just wistful thinking? We've put that question under the microscope, too, and the answer is not all science fiction. In fact, as well as proving to be a serious line of scientific and philosophical inquiry, it threw up timeless matters of love and loss – a physicist whose father died when he was ten became a time travel expert so he could go back and see his dad again.

Journalists across the newsrooms of *The Age* and *The Sydney Morning Herald* have been sharing their expertise in 'explainer' articles for three years now. A handful of the most colourful went into our first anthology, *What's It Like to Be Chased by a Cassowary?* in 2020. We aim to give our readers clear and interesting explanations of concepts, buzzwords or events in the news and in life – to get to the guts of an issue without sacrificing the nuance. If you've enjoyed getting to grips with a topic through reading one of our explainers and you've come away with some pithy facts to share with family and friends, then we've done our job.

Ballet dancers are, traditionally, not as fit or strong as you might think. Tanning is the body's stress response to damaged DNA. Long before it arrived in Canberra, the word 'faction' was used for chariot-racing teams in ancient Rome. The Earth's seas are a highway for hulking ships that navigate choke points, geopolitical hotspots and, sometimes, even pirates to bring us half the stuff in our homes. France is the second biggest hip-hop market in the world. And, here's a fact for a romantic first date: sniffing a prospective lover is one of the best ways to tell if they're right for you.

Once again, the *Herald*'s creative director, Dionne Gain, has brought visual style and wit to our explanations with her illustrations; and the twin investigative and storytelling talents of explainer reporter Sherryn Groch are on show, with some cracking unpacking of why sharks bite, how a cyberwar might play out, how to use gender pronouns, how to solve homelessness (there is an answer to this problem we are so often told is insoluble) and more. Among the feedback Groch has received for these pieces was this from Glenn Kent, whose quest for a roof over his head is a central story in 'How do you solve homelessness?' Glenn told Sherryn, 'You listened so I spoke from the heart and you got it just right.'

As the national explainer editor, I've interpreted my brief broadly – some of the pieces in this book are not what we'd call hard news, but they're not exactly soft either. The wider world and its politics appear: how will the next Dalai Lama be chosen? Why do Russians get poisoned? How does the Communist Party of China work? And science, its importance so indelibly highlighted by the COVID pandemic, is intrinsic to our explainers on art forgery, love at first sight, elite swimming, psychedelic drugs as medical treatment, the carbon-drenched richness of our sea meadows and how your cat and dog see the world.

But, no matter the topic, all of these explainers are really about being human, how we live, our own personal travels in time. You'll see that writer Jewel Topsfield has explored 'What is it like to live with dementia?' through the voices of people who live with one of its many forms and those who care for them – an increasing number of Australians. And, far from being a depressing chronicle of the road downhill, in 'How do we age?' Sophie Aubrey blows away some cobwebby ideas, with one octogenarian declaring he wouldn't time travel back to his youth even if he could. Even our younger readers are advised to start warming up for their later years now – Aubrey says her extensive research led her to conclude that 'ageing is a gift we should embrace and plan for early on'.

Where do the ideas for such explainers come from? From readers, reporters, colleagues and me. From the mysterious hinterlands of our minds, niggling questions emerge for all of us. You can shrug or you can seek insights into whatever it is that's puzzling you. Are you curious about the same things we are? We never know until an explainer is published. When you've read this book, we hope your reaction will be similar to that of one wag who, on finishing Groch's time travel explainer, commented, 'As Spock would say, "Fascinating."'

1

WHERE DID THE AUSTRALIAN ACCENT COME FROM?

We all speak with an accent, but what is it that makes us sound Australian?

Karl Quinn

W̲e all know an Australian accent when we hear one: the single word 'g'day' can be enough to signal that the speaker is either (a) Australian or (b) some Hollywood actor pretending to be Australian. And depending on how they say 'g'day', we might also be able to guess something about their background (or acting skills).

A slowly drawled 'g'day' and chances are they're from the country. A short, sharp 'g'day' and they're probably a city dweller. An overly enunciated 'good day' and they likely spent summers on Daddy's yacht.

But where did this accent come from? What are its key characteristics? And is there only one Australian accent, a couple, or many?

I don't have an accent . . . do I?

The first thing to know about accents is that we all have one, even if speakers of what socio-linguists call 'general Australian English' might be tempted to think only other people do. As the Linguistic Society of America helpfully puts it, 'broadly stated, your accent is the way you sound when you speak'. There are two different kinds of accent, the society adds: the 'foreign' iteration where 'a person speaks one language using some of the rules or sounds of another one'; and the local one, 'the way a group of people speak their native language, [which] is determined by where they live and what social groups they belong to'.

> *The first thing to know about accents*
> *is that we all have one.*

If you live in Australia and speak English, you will fall into one of these groups. If you were born here, or born overseas and educated here, you will likely fall into the local group – though the scope for variation within that group is pretty wide. If you were born overseas but arrived as an adult, you might speak English either as a first language (if you're from Britain, say, or Canada) or as a second language (if you were born in China or Vietnam, for instance). Either way, you will be classed as a foreign speaker insofar as accent is concerned.

So, there's more than one kind of Australian English?

Too bloody right there is, cobber.

Our understanding of the Australian accent owes much to two academics at the University of Sydney, Alex Mitchell and Arthur Delbridge, who began in the late 1950s to record and analyse the speech of about 7000 teenagers. Published in 1965 as *The Speech of Australian Adolescents*, their work discerned three clear strands of Australian English: broad (your classic *Dad and Dave*–style 'strine'); general (which most people spoke, whether in the cities or in the bush); and cultivated (the posh version spoken in the more well-to-do pockets of the major cities).

The vast majority of Australians, then and now, tend to speak a version of general Australian English with only minor regional variations – not enough to amount to dialects, the regionally distinct forms complete with vocabularies of their own, as spoken in different parts of Britain and the United States.

But despite the drift towards homogeneity, linguists have also begun to embrace the idea that Australian English encompasses not just broad, general and cultivated forms – collectively referred to as 'mainstream' Australian English – but also Aboriginal and ethnocultural accents (and where they are accompanied by distinct turns of phrase and patterns of speech, these are sometimes called ethnolects, for ethnic dialects).

Is our accent 'proudly Australian' then?

The Australian English accent has come in and out of favour – we certainly haven't always been happy to sound Aussie. From the end of the nineteenth century until the 1960s, there was a push in Australian education towards a standard accent, based on the RP (received pronunciation) taught in the British public-school system. Implicit in that was the idea that there was a correct way of speaking, and that anything non-standard was a degraded form.

But the notion of a standard to which we should all aspire has fallen from favour as part of a general shift away from judging the way people speak to simply recording it. 'All dialects reflect Australian identity but, in addition, reveal the cultural affiliation of the speaker,' says Felicity Cox, a professor in linguistics at Macquarie University. Indeed, Cox says that Australian English may soon come to function not as a descriptor but as an overarching term 'embracing these various dialectal types rather than excluding minority forms'.

'Such a modification to the traditional concept of Australian English,' she adds, 'will help capture the linguistic landscape of the changing Australian culture.'

So, Australian English is this ever-evolving thing?

Yep. As Debbie Loakes, a postdoctoral fellow in languages and linguistics at the University of Melbourne, puts it: 'There have been some substantial changes in the last 20 years to Australian English – but it is also true that all languages change, and that's one of the main things you can rely on with language.' (To get a sense of that change, check out the excellent body of recordings of Australian speakers from the past at goto.mq/australianvoices and search for Australian Ancestors Project.)

The way we speak isn't just about accent, though; it's also about the words and turns of phrase we use. As Bruce Moore puts it in his book *Speaking Our Language: The Story of Australian English*, 'The accent in which . . . words are spoken is an essential part of their meaning.'

This is the stuff of idiom, slang and dialect, and it is informed by region, class, social cohort – and the times. For instance, teenagers today will casually drop terms such as 'bae' (slang for 'babe' and an acronym for 'before anyone else', to denote a best friend) or 'lol' (laugh out loud – not to be confused with Grandma's Christmas card sign-off, 'lots of love'). These terms derive from text messaging and social media but have infiltrated the realm of the spoken word. And while some older speakers might decry them as corruptions of the language, they are simply part of its constant evolution, no different than the 'strike me pink' (an expression of surprise, roughly equivalent to 'crikey') or 'avagoyamug' (a somewhat aggressive exhortation to maybe try a bit harder) of earlier times.

But where did the Australian accent come from?

In a word, England. In a few more words, the south-east of England. In a few *more* words, the kids of the kids of the settlers who arrived with the First Fleet, many of whom came from south-east England.

In 1788, 732 convicts (by Governor Arthur Phillip's count) and a few hundred jailers, administrators and family members settled in Sydney. Many of them came from – or, at least, had been sentenced in – London, but there were a good number from elsewhere in Britain: York in the north, Coventry in the Midlands, Exeter and Dorset in the south-west, Maidstone in Kent and plenty of other places besides (including Ireland, Scotland and Wales). If you're at all familiar with the range of accents and dialects spoken in the British Isles you'll realise the Rocks must have sounded a bit like Babel, with a cacophony of different regional tongues and idioms struggling for common ground or dominance.

For a time, it was assumed that the Australian accent was little more than a version of Cockney, the dialect traditionally spoken by the working-class inhabitants of London's East End. But as Bruce Moore writes, 'It is now clear that the Australian accent was not "transported" holus bolus to Australia from some part of Britain, but that it developed in Australia.'

The Australian accent emerged in three stages, he suggests. The first involved some voluntary modification of dialect and idiom as the settlers sought to make themselves better understood. The second was an acceleration of that trend as their children strove to sound less like their parents and more like one another (though traces of their parents' accents necessarily remained). And the third emerged as the children of those children unconsciously selected out yet more regional variation from the Old Country, to arrive at a relatively stable local dialect.

Because there are no recordings, this account necessarily involves a good deal of hypothesising and debate. But it is generally accepted that the process known as 'levelling' was completed by the 1840s, and the mode of speech that emerged then spread from Sydney throughout the rest of the country as the settler population dispersed.

As Moore writes: 'The accent that is established in this second generation will be passed down to future generations, and it will be almost impossible for any later groups of migrants to influence it. The "foundation accent" has now been established.'

But what did it sound like?

Some linguists have argued that the foundation accent was a variant of broad Australian but many are convinced it was, in fact, much more like the general Australian accent most of us speak today. 'It would sound different to the modern accent because accents are continuously changing,' says Cox. 'But it was the precursor of the modern-day Australian accent.'

Visitors from England were struck by the pleasing uniformity of this 'native' tongue. There were, of course, many other native tongues – at least 250 by recent estimates – spoken by First Nations people long before white settlers arrived, and some of their words –

For a time, it was assumed that the Australian accent was little more than a version of Cockney.

kangaroo, galah and goanna, to name a few – found their way into Australian English, although Aboriginal languages appear to have had little impact on the settler accent.

Chances are those early speakers of Australian English had already uttered the distinctive vowels and diphthongs (a combination of two or more vowels, such as 'oi' or 'ai') that are the clearest markers of our accent. And if so, it was deemed an improvement on the regional British dialects that preceded it.

Moore's analysis of the historical record shows that 'the typical descriptions of the Australian accent for most of the nineteenth century are not adversely critical'. He cites an account from 1822 in which visiting Englishman James Dixon claims the children of the colony 'speak a better language, purer, more harmonious than is generally the case in most parts of England. The amalgamation of such various dialects assembled together seems to improve the mode of articulating the words.'

Our tendency to drawl was not really identified as a problem until the 1880s. But by the first half of the twentieth century, we were in full shame mode, with our accent regarded, Moore writes, 'as impure, ugly, and sub-standard'.

So, beauty is in the ear of the beholder?

'Ken oath, it is.

Take the way we pronounce the word 'dance'. Today, we tend to assume that someone who says 'darnce' (imagine Cate Blanchett saying it) is either posh, pretentious or from Adelaide. (It is impossible to say definitively, but it is possible that 'darnce' and 'charnce' may have gained in popularity in Adelaide as a way of distinguishing it as a free settlement – a place where people were subjects of England rather than subject to it.)

But it was not always so.

In 1791, John Walker's influential *Critical Pronouncing Dictionary* made it clear that pronouncing the vowel in words like 'plant' with a long vowel (as in 'car') 'borders very closely on vulgarity'. By that measure, saying 'darnce' might have been deemed exceedingly common back then.

If it's true that both the broad and the cultivated accents really did emerge after general Australian, it's likely they did so in response to our relations with the mother country. 'They were both aberrations driven by social elements,' says Moore. 'Think of them as conflicts with Britain that were played out verbally.'

Cox says there is evidence to support the idea that broad Australian gained ground during the First World War, spoken in the trenches by Diggers as a marker of difference from, and disdain for, the British officers who commanded them. Cultivated Australian, by contrast, was embraced by some others as an expression of loyalty to Queen and country.

Jenny Price conducted her PhD on the changing speech patterns of Australian newsreaders between 1951 and 2005. Examining the recorded speech of 80 newsreaders, she detected a gradual shift over time from cultivated towards general Australian. 'My starting point was this old-fashioned accent that came from the UK and is linked to our colonial origins,' she says, referring to that particularly rarefied voice heard on old newsreel accounts of, say, the arrival of Charles Kingsford Smith's plane in Australia, or Bradman's latest knock ('the Orstrellyuns are getting in plenty of precktice by way of metches for the Sheffield Shield').

The arrival of radio in Britain in the 1920s, Price says, 'presented fantastic opportunities for linguistic education and influence', an opportunity to standardise speech and eradicate stigmatised regional accents. 'So the BBC deliberately defined what they wanted their accent to be, and they settled upon general RP.'

When the new technology arrived in Australia, where our general accent was similarly criticised 'for being careless, lazy and excessively nasal', we followed suit. Price cites one account of the typical broadcaster voice of the era as 'presenting a news report as though they were reading the Ten Commandments with Divine permission'.

Television followed much the same template when it arrived in 1956, but later advances in technology were crucial in turning the tide away from the cultivated accent. Smaller, lighter cameras and tape recorders allowed reporters and crews to get out of the studio and into the street, and to bring the voice of the ordinary Australian into the living rooms of other ordinary Australians. And those ordinary Australians soon realised they weren't alone in not speaking like a BBC (or ABC) announcer.

A rising sense of national pride from the late 1960s further reinforced the idea that there was nothing to be ashamed of in sounding like an Aussie. There was a surge in popularity of what you might call a 'performed' version of the broad accent in popular culture from that time right through to the late 1980s, from Bazza McKenzie ('Don't come the raw prawn with me!') through Ted Bullpitt ('You're not taking the Kingswood!') to Crocodile Dundee ('That's not a knife') on screen; Dave Warner ('Just a Suburban Boy') and The Angels (their anthemic 'Am I Ever Gonna See Your Face Again?' famously drawing the audience response, 'No way, get fucked – fuck off!') in song; and John O'Grady (*They're a Weird Mob*, under the pen name Nino Culotta) and David Williamson (*Don's Party, The Club*) on the page and stage.

> *A rising sense of national pride from the late 1960s
> further reinforced the idea that there was nothing
> to be ashamed of in sounding like an Aussie.*

Which is not to say the broad accent didn't also exist in the real world. It did, and it served as a marker of both class (lower) and location (generally regional). These days, it mostly survives only as a geographical marker. Outside of places such as 'Brahton' (aka the Melbourne bayside suburb of Brighton, stomping ground of Prue and Trude from the TV comedy *Kath and Kim*), the cultivated accent has also largely disappeared. 'And that suggests to me that the old conflict with Britain has been almost completely resolved in Australian society,' observes Moore.

What does this general Australian accent actually sound like?

Australian English is all about vowel movements. 'And the thing is, vowels operate in a system,' says Cox. 'So, when one vowel changes it can impact on the entire system.'

To illustrate, she offers an example from New Zealand English. 'The vowel in their "pat" sounds like the vowel in our "pet", so when they say "pet" it sounds to Australians like "pit". It's like dominoes – one thing changes and other things have to change as well. And in New Zealand, the vowel sound in "pit" had nowhere else to go, so it had to go to a central sound, like "puht".' (Linguists call this type of unstressed central vowel sound a 'schwa'.)

In Australia, Cox says, 'what's happened in the last 30 years is that the vowel sound in "pat" is now really low in young people, so "laptop" sounds like "luptop". So the next one, the "e", comes down too, so "pet" sounds much more like "pat".'

Australian English, she adds, has a very high number of vowel sounds – 19, in fact. 'But there's change going on all the time.'

Anna McCrossin-Owen, an accent coach whose clients include actors, radio professionals and immigrants, says we have some very long vowel sounds 'and we like to emphasise a lot of our words'. Compared to American accents, which she says are particularly 'muscular' (meaning they involve considerable use of the facial muscles), she says, 'Australian articulation requires a more subtle awareness of movement, and I think this is what can make it difficult for Americans and other speakers.'

The biggest challenge for a non-Australian speaker trying to emulate our accent, notes opera singer and speech therapist Sarah Lobegeiger de Rodriguez, is probably our diphthongs. 'Any time we have two vowels that sync together in a word like "rain", "boy", "how" the challenge is finding the duration because Australian vowels tend to have a very strong, long duration on the first vowel,' she says.

While that is particularly pronounced in the broad accent, she says we all tend to deliberately change our accent depending on circumstance and company (it's what's known as 'speech accommodation', and it's common across speakers of all languages).

'You could have a QC, say, who's very proper in court but then he could be at the pub with his mates saying, "Howzitgarn?" When we communicate, we've got neuro-linguistic processes going, with how we move our jaw and mouth and muscle movements; we've got the cultural speech patterns from early childhood in the home and school and colleagues and friends; but then we also have a pragmatic system, what words will I use for my message, giving eye contact, using hand gesture or perhaps sarcasm.'

Some other markers of the modern general Australian accent are more seasonal in nature: upspeak (where we go up at the end of sentences) is in decline; vocal fry (where the ends of sentences are under-aspirated, leading to a kind of drawn-out croaking sound) is gaining ground in both male and female speakers; and the phenomenon of 'L-vocalisation' – in which 'noodle' is pronounced 'nooduh' – is on the march from Adelaide, where it appears to have first emerged.

So, do we all speak general Australian English now?

Not quite. Although the trend is towards a homogenised general accent, linguists are increasingly looking to codify the wide variety of other accents found in modern Australian English. 'There are changes occurring now, really over the last 10 to 20 years, where sub-dialects and sociolects are strongly established,' says McCrossin-Owen. 'For example, first-, second- and even third-generation European accents such as Greek, Lebanese, Turkish and Italian all have their own sub-dialects.'

There are also clear markers of difference in the way Australian-born English speakers from Asian backgrounds talk; likewise with the descendants of people from the Middle East, Africa and the Indian subcontinent, whose accents 'establish themselves through generational imitation of the family accent blended with the social accents that are heard', says McCrossin-Owen. Aboriginal English also has its own distinctive patterns and sounds.

Where once upon a time these variations might have been seen as failings to speak standard English, they are increasingly celebrated by socio-linguists as part of the rich tapestry of modern English in Australia.

'There's nothing inherently correct or incorrect about language – it's all about the use to which it is put in its community,' says Cox.

That said, stigmas do still exist. Just ask Sarah Lobegeiger de Rodriguez. Many of her clients are fluent speakers from countries where English is an official language but they still struggle to make themselves understood here. As a consequence, they lose confidence, become reticent about speaking, have trouble finding employment, and the problem only intensifies. Our perception of accent, in her view, is political – and effort is needed on the part of listeners as well as speakers to achieve an equitable outcome. 'Accent is always relative to the listener; there's no such thing as a strong accent,' she says. 'I tell clients, "You've got a relevant voice. You are the sound of modern Australian English more than I can ever be" because we're not a nation of monolinguals.'

And that, says Cox, is something we should all be celebrating.

'We haven't seen this type of diversity before, ever,' she says. 'We're seeing the birth of new varieties of English, in Australia and across the world. And that's really exciting.'

2

HOW DO YOU FAKE A MASTERPIECE?

It's difficult to get away with faking an artwork, but not impossible. How many fakes are out there, and how are they unmasked?

Melanie Kembrey

I've always wanted to own a Picasso. As luck would have it, there's one for sale online. It's a hand-signed etching with stamps of authenticity from the Musée d'Orsay in Paris and Christie's New York. It costs $10,000. The listing says it's a rare opportunity. I'd either be a fool if I did or a fool if I didn't – and I can't quite figure out which kind of fool I'd rather be. So I run my potential acquisition by one of the world's leading experts in art crime.

'I would suggest,' Professor Duncan Chappell replies, 'that investing in it would be a swift way to lose the family silver.'

There are a couple of giveaways. The etching would be sent from a small outer Sydney suburb, which is 'hardly the sort of spot you would expect a stray Picasso print to be sourced from'. Even more troubling, according to Chappell, is that the Musée d'Orsay and Christie's do not usually provide certificates of authenticity.

So, it seems my Picasso is a con. But while the etching is fake, the issue of art fraud is real. Chappell, who co-edited the *Palgrave Handbook on Art Crime* and is a former director of the Australian Institute of Criminology, says a conservative estimate is that 10 per cent of artworks, in Australia and internationally, are of questionable origin. Art historian Thomas Hoving, the former director of New York's Metropolitan Museum of Art, claims that up to 40 per cent of thousands of works he examines could be considered fakes, while Sharon Flescher, the executive director of the International Foundation of Art Research, has reported that close to 80 per cent of works submitted to the organisation were not by the attributed artist.

10 per cent of artworks, in Australia and internationally, are of questionable origin.

There are plenty of whispers and anecdotes about, and unofficial blacklists of, forged works – in Australia the market is small, after all – but the cases that do gain public notoriety usually involve big names, big galleries, big coin and big courts.

For artists, they can be a huge drain on money, time and spirit. The late Robert Dickerson, one of Australia's most famous figurative painters, went to court to have fakes of his work destroyed. He won, but his son Sam says, 'He found it depressing and deflating. It just distressed my dad so much.'

So what does it take to fake a work of art? And what happens if you get caught?

What is a fake?

Art fraud is one of three key types of art crime, alongside art theft and cultural heritage offences, which includes antiquities looted from conflict sites. All three hinge on one word: provenance. From the French *provenir*, meaning to come from, provenance is an artwork's biography, including authorship and ownership. It can include receipts of purchase and sales, exhibition catalogues, notes and photographs of works in progress. Chappell says the terrain of art fraud is 'cluttered with meanings that may cause occasional confusion', with terms including copy, forgery and fake used loosely. But they are distinct in practice.

Copying an artwork, or painting in the style of an artist, is not a crime. Artists have long honed their craft by copying the masters. There are dozens of imitations of Leonardo da Vinci's *Mona Lisa*, including the Prado *Mona Lisa,* which is believed to have been painted by one of his apprentices at the same time and in the same workshop as the version that hangs in the Louvre. Chappell says under English common law the term 'forgery' does not apply to works of art but to the forging of documents or writing – and it is the false signature added to an artwork that qualifies as the forgery.

The term 'fake', however, 'implies both the work in question is not authentic, and that it has been intentionally produced to deceive', says Chappell. A fake involves a physical element (the painting that is not what it appears to be) and the mental element (the intention to deceive). Faking an artwork usually falls under the category of fraud rather than forgery, although selling a fraudulent artwork might involve forgery, including doctoring documents of authenticity.

Yet in common parlance, those who create fake works are often referred to as forgers. Biographies of Michelangelo suggest the great Italian artist dabbled in fakery early in his career, doctoring a statue to appear ancient and selling it to a cardinal who collected Roman antiquities. Professor Alexander Nagel, from New York University's Institute of Fine Arts, has highlighted how the concept of forgery developed parallel to the art market in the West, from around the year 1500, and before then originals and copies were happily interchangeable. Since then, artists including Andy Warhol have mocked the concept of individual authorship, blurring the fake and the genuine in their practices, and making authentication of their works challenging.

What does faking a masterpiece involve?

A good fake requires artistic talent and expert knowledge. The British artist John Myatt perpetrated one of the biggest art frauds of the twentieth century by faking works by Henri Matisse and other major artists – which were sold by his co-conspirator John Drewe, who forged documents to convince buyers they were genuine – before they were both imprisoned in England in 1999. Myatt later told *The Guardian* that he wanted to be hypnotised by the artist whose work he was faking. 'I like to know everything – where he was, what he was doing, what his relationship was like with his wife when he was painting.' Fraudsters such as Myatt replicate the nuance of an artist's practice – style, signature, the application of brushstrokes – and source materials and frames from the correct time period. German fraudster Wolfgang Beltracchi made many millions of dollars before he was famously caught out using titanium white paint in a Heinrich Campendonk knock-off meant to be from 1914 – the pigment did not exist at the time.

*A good fake requires artistic talent
and expert knowledge.*

Some fraudsters seek to avoid attracting attention by selling their work for modest prices. They will look to copy smaller pieces by lesser-known artists prolific in their output and liberal in the distribution of their art, or by artists whose biographical details are sketchy or disrupted by illness or a war – because this leaves room to concoct a convincing explanation for why they can't prove a work is genuine. It's also easier if the artist is not living, if their estate is not closely guarded by relatives and if there is no catalogue raisonné, a document detailing all known works by an artist.

That's not to say fraudsters don't go big. The art world was rocked by false works purporting to be by major abstract expressionists being sold as real by Manhattan gallery Knoedler & Co., which closed in 2011 ahead of major lawsuits from collectors who had bought the pieces. The $80 million fraud, the subject of a Netflix documentary *Made You Look: A True Story About Fake Art*, saw dealer Glafira Rosales sell more than 60 works to the gallery, claiming she was given them by a collector called Mr X. The gallery, one of the oldest in America, sold the paintings for more than 15 years, turning a significant profit. The painter turned out to be Chinese immigrant Pei-Shen Qian, who made the fakes in his Queens garage, which when raided by the FBI contained 'books on abstract expressionist artists and their techniques;

auction catalogues containing works by famous American abstract expressionist artists; paints, brushes, canvases and other materials, including an envelope of old nails marked "Mark Rothko".' Rosales pleaded guilty to multiple charges but Qian has maintained he was the victim of a 'very big misunderstanding' – he claims he never intended the works to be sold as the real thing, and never received more than $7000 for a painting. He was indicted in the United States on charges of wire fraud, conspiracy to commit wire fraud and lying to the FBI in relation to the fraud, but now lives in China. 'I made a knife to cut fruit,' he told Bloomberg News. 'But if others use it to kill, blaming me is unfair.'

How are fakes discovered?

The art market is on high alert for art fraud. Bonhams Australia director Merryn Schriever says auction houses take a 'very cautious approach' in valuing, cataloguing and presenting works to market. 'We always like, where possible, to really give pictures or objects back their identity if they have been lost,' says Schriever. 'We do it for everything we have in the catalogue. If they are not diligently executed and researched they can become liabilities on your ledger.'

Auction houses, agents and sellers can be accountable for false representations about an artwork that results in losses for buyers. In 2014, the NSW Supreme Court ordered Christie's to cover the bulk of $120,000 in damages after they sold a work purported to be by Albert Tucker. Sydney barrister Louise McBride bought *Faun and Parrot* for $85,000 and hung it in her home for more than a decade until she tried to sell it at auction, and doubts about its authenticity emerged. McBride took Christie's, and the fine art dealership that consigned the painting to Christie's as well as her art adviser, to court and won. The court heard that the parties did not know the painting was a fake at the time of the auction. Soon after, however, Christie's was advised by a group of art experts of 'real concerns' about the work's authenticity, and did not inform

McBride. Justice Patricia Bergin ruled that 'Christie's conduct was commercially reprehensible and unconscionable' and the auction house had an obligation to correct its representation that there was 'no doubt' the painting was by Tucker.

The need to establish provenance is a process that turns art experts into art detectives as they scour archives and consult authorities and estates to establish a paper trail of authenticity. Galleries, auction houses, collectors and dealers also have science at their disposal, including chemical analysis, radiography and microscopy. Chappell says these processes can help establish whether the pigments and materials used in a work are consistent with those available at the time it was purportedly made, and allow for a thorough stylistic examination to establish if, for example, the subtleties of a brushstroke are consistent with the artist's style.

And the sleuthing can end up 'reframing' the story behind an artwork. The painting *Head of a Man* (1886), for example, was thought to be by Vincent van Gogh when the National Gallery of Victoria bought it in 1940. In 2007 it was sent to the Van Gogh Museum in Amsterdam for analysis by X-ray (which can show, for example, how a painting is constructed) and under the microscope (which can show, say, exactly what paints were used), as well as for appraisal by connoisseurs (who compare the work under a microscope, too, with other works from a similar time and place). The painting was found to be from van Gogh's time – but was not by the famed artist at all. The Amsterdam museum also found that the painting had belonged to a collector, known only as 'S', from Berlin; NGV research in 2011 confirmed that the collector was Jewish manufacturer Richard Semmel. By 2012, the NGV had discovered through Semmel's heirs that he had been forced to sell the painting as a result of Nazi persecution. The gallery returned *Head of a Man* to Semmel's family, who have loaned it back for display – and the story is outlined in a plaque that hangs beside the painting.

Some works elude attribution. A recent case was the painting *Cityscape* (1950s) that came to Bonhams with little background from a collector. Schriever, who had a strong suspicion about the identity of the artist, showed the work to curators, conservators and consultants, as well as descendants of the attributed artist, and undertook a full scientific analysis. But Schriever says they hit a dead end and the work was sold as 'artist unknown (twentieth century)' for $25,830 in April 2021. The auction house Menzies, too, revealed the industry's caution when it withdrew two small paintings believed to be by Australian surrealist painter Peter Booth before auction in April to allow more research into their provenance.

> *It pays to look a little longer and ask a few harder questions before investing in art.*

Galleries have also been rocked by provenance scandals, although mainly relating to stolen antiquities. The highest-profile case was that of a 900-year-old Dancing Shiva bronze that the National Gallery in Canberra bought in 2008 from New York antiquities dealer Subhash Kapoor, who was arrested and charged in India for allegedly running an international smuggling racket. The gallery returned the statue to India in 2014 after it was found to have been stolen from a temple and illegally trafficked. In July 2021, it returned 14 more works bought from Kapoor between 1989 and 2009. Another 12 were being investigated. 'I think it's very important to be conservative and careful and do your due diligence across the board and we, unfortunately, learnt the hard way a very valuable lesson,' says NGA director Nick Mitzevich, who was appointed to the position in 2018. He says there will be 'zero tolerance' for gaps in provenance.

Mitzevich likes to think there are no fake masterpieces hanging in the NGA but anyone is welcome to contact the gallery if they have doubts. The gallery's provenance department looks at four elements to establish an artwork's authenticity: science, published and unpublished records, experts, and artist estates and foundations. 'We run through all the exhaustive work before we come to a conclusion. If we identify any inconsistencies, we will not recommend acquisition.' Yet provenance can be a moveable feast, as highlighted by the NGV's *A Monk with a Book*, circa 1550, which was attributed to Titian in 1924, faced doubts in 1971 and in the 1980s, but in more recent years has been accepted as being by the Venetian painter.

Have there been court cases in Australia involving art fraud?

The highest-value case of alleged art fraud in Australian history involved three paintings in the style of Brett Whiteley that were sold or attempted to be sold for a total of $4.5 million. Arts journalist Gabriella Coslovich, who sat through the five-week trial and wrote the Walkley Award–winning book *Whiteley on Trial* about the case, says that no more than one art fraud case a decade seems to reach the criminal courts in Australia and the lasting effect of the Whiteley case has been a 'feeling of dismay and reticence' to pursue a criminal conviction for art fraud.

Before their convictions were sensationally overturned, art dealer Peter Gant and art conservator Aman Siddique were found guilty of three counts each of obtaining and attempting to obtain financial advantage by deception by a jury in the Supreme Court of Victoria in 2016. The prosecution argued that Siddique created the alleged fakes, two of which were sold as authentic Whiteleys, one for $2.5 million in 2007, the other for $1.1 million in 2009 (the third one did not sell). Gant and Siddique pleaded not guilty. Gant was given a five-year prison term and Siddique three years.

In a rare decision, Justice Michael Croucher stayed the

sentences until Gant and Siddique's appeal was heard. Their convictions were quashed in 2017 in the Victorian Court of Appeal when the prosecution itself conceded there was a 'significant possibility that innocent men have been convicted, and each of them should accordingly be acquitted'. Justices Mark Weinberg, Phillip Priest and Stephen McLeish said they were not equipped to make a finding about whether the paintings were real or fake. In delivering the ruling, Justice Weinberg said, 'It is sometimes said that juries "always get it right". Sadly, in this particular instance, that seems not to have been so.' He described the case as 'a rare and almost unique instance of the system having failed'.

There are only two known cases in Australia in recent years that have seen convictions for art fraud. One involved husband and wife Pamela and Ivan Liberto, who made more than $300,000 faking and selling paintings they said were by Indigenous artist Rover Thomas. Scientific analysis established the works were not authentic, and police found paints, art catalogues and works in progress when they arrested the couple at their house.

In the County Court of Victoria, the Libertos were found guilty of charges relating to attempting to and obtaining money by deception and were sentenced to three years in jail in 2007, of which two years and three months were suspended. Art dealer John O'Loughlin pleaded guilty in the NSW District Court in 2000 and was given three years' probation for selling paintings – including to the Art Gallery of New South Wales and the Museum of Contemporary Art – that he claimed were by renowned Indigenous artist Clifford Possum Tjapaltjarri.

Civil law actions can also be taken against those involved in perpetuating art fraud. Chappell says the benefits of civil court is that the standard of proof is lower than in criminal proceedings, so victims have a greater chance of recovering funds and of influencing the process. Australian artists Charles Blackman and

Robert Dickerson sued Peter Gant in 2010 for breaches of the Fair Trading Act for selling fake works carrying their names. 'It makes me look stupid,' Dickerson told the court. 'It's a bad drawing, terribly bad.' Gant argued that the works were authentic. Justice Peter Vickery ruled the works were 'fakes masquerading as the genuine article' but made no finding against Gant. Blackman and Dickerson were awarded damages and won the right to destroy the fakes. They burnt them on a bonfire. 'I feel very happy about that because I am sick of these fakes floating around,' said Dickerson, then 86, of the ruling. 'It has been a very long case but it's been worthwhile. It hasn't been a waste of time.'

But civil cases are costly and time-consuming, and there's no guarantee you'll receive the money owed. Sam Dickerson says his father, who died in 2015, was determined but deflated. 'My dad wasn't young when he took it on. He didn't like suiting up and going to courts. To have to actually prove you didn't do something – the crux of that argument in that case was to have to explain it wasn't your work – was difficult.'

Why are there so few cases in Australia?

First, we don't actually know the extent of the problem. There are no official statistics to establish trends in art crime because there is no specific crime category under Australian state or federal law, nor an art crime squad. There is no national register of fakes or set rules for how the market deals with problematic artworks, unlike in France where fakes can be destroyed. Chappell says virtually all art fraud is 'dark crime', not to be found in reported crime figures. 'It seems that the most reasonable conclusion is that when it comes to either numbers of frauds, or numbers of offenders, the real figures are not likely to be large,' Chappell says. 'Having said that, this does not mean that there are but a few false works available on the market. The output of even a few of those creating fraudulent work has been huge.' So, while the number of offenders may be relatively low, the 'cumulative effect on the market has been quite large'.

Second, the Australian art industry has only a dozen or so major auction houses and dealers, so the opportunity for a fraudster to repeatedly get fakes into the market is limited, but where it does exist, shame and silence about fake art perpetuate the potential for it to circulate. No one likes to be tricked. The high-stakes world of the art market is one of power, pride and prestige, and it can be less painful for a collector to quietly take down a dubious painting and putty over the nail holes, or leave it hanging, rather than admit to being sold a dud. Victims could also be motivated to stay quiet, because acknowledging a masterpiece as fake can render a major financial investment worthless. They could instead return the work to the market with its original attribution in place. The identities of both owners and buyers are protected from public view; a work will have been said to have entered a 'private collection'. 'I've definitely heard of stories where a collector knows they've got something weird on their hands and they just put it aside, they don't want to go there, they don't want to know,' Coslovich says. Dealers privately settle with buyers if an artwork is found to be questionable to avoid the public relations disaster or because they are obliged to via their sale contract. They can also be reluctant to discuss the issue out of fear it would rattle confidence in the market.

The scarcity of cases that reach court means there is little experience when it comes to dealing with art fraud. Coslovich says the Whiteley case demonstrated a lack of appreciation for the role of connoisseurship, with expert evidence about the alleged fakes given little weight. Connoisseurship is about knowledge and instinct, she says, which can be hard to reconcile with the criminal court's burden of proof – the requirement for an alleged crime to be proven beyond reasonable doubt. 'There was little regard for the evidence given by the art experts and their discussion about the points of identification of an artist's work. If you've looked

It will become harder to get away with art fraud as technology advances.

at lots of Whiteleys you become familiar with the artist's style, or what is known as the artist's "hand".' Coslovich says better education of police, prosecution and members of the judiciary is key to changing how art crime is managed.

In any case, it will become harder to get away with art fraud as technology advances. One company is even looking at how artists might tag their work with a unique synthetic DNA sample. At the moment, while scientific analysis of paintings has become largely de rigueur, Chappell says there remain some 'fuddy-duddy' attitudes towards it by older galleries, auction houses and collectors. 'I think it's conservative views, perhaps, in the art market, and also a false sense of parsimony to not pay out the fees which you might not have to otherwise, to get a scientific audit of the artwork,' he says. 'I think if I were buying, I would get a scientific assessment made.'

Why does art fraud matter?

It can be easy to dismiss art fraud as unimportant – why should we care that a person with too much money spent too much of it on a fake masterpiece? Ann Freedman, who ran the scandal-ridden Knoedler Gallery, told *The Art Newspaper* in 2016 she was 'terribly sorry' for anybody who said they had been 'hurt or damaged' through buying fake masterpieces, 'But let me be clear, this is [about] works of art. I didn't slay anybody's first-born. We have to have some perspective on suffering.'

While maintaining perspective is important, art fraud undermines the integrity of art and destabilises the art industry more broadly. It also causes suffering to the artist who is the subject of the fake, impacting on their reputation and sales. 'In my opinion, the worst part of those things being in the market is the harm done to the legacy of those artists,' Schriever says. 'I find that when fakes are out there in the market, if and when we see them it besmirches the artist themselves.'

Many of the future issues related to art fraud in Australia will be tested in First Nations art and craft. Research by the Arts Law Centre of Australia found that as much as 80 per cent of First Nations art and craft in tourist shops is not authentic, including items falsely attributed to Indigenous artists and designs reproduced without permission. This is a particularly pernicious area of art fraud as it is widespread and hard to control through the supply chain, affects economic opportunities for artists and undermines the integrity of First Nations culture. A parliamentary inquiry reported in 2018 that 'First Nations artists and their communities feel completely disrespected and cheated by what is going on at the moment . . . This unacceptable misappropriation of First Nations cultures cannot be allowed to continue unchecked.' The government is considering laws to stop the proliferation of fake art and souvenirs, which could have broader ramifications for fraud in the art world. While the fine art market is less affected, as auction houses and galleries have firm guidelines for Indigenous art, the industry is still working out how to avoid unethical practices.

Sam Dickerson, who runs the Dickerson Gallery Sydney, says he is sure there are still a handful of forgeries of his father's work out there, but he hasn't seen one since the civil case. In the past, he detected a dozen or so fakes said to have been by his father – the giveaways included the type of paper, certain ways his father mounted his work, his signature. 'People always try to improve what my dad did, to try to make it look better but it then looks too nice, too finished, too angular,' he says. While going to court was heart-breaking for his father, the judgment 'lit a fire under everyone to be a bit more aware'. 'It certainly does show that people are watching. For us that was the most important thing, to say that we are here.'

3

WHY DO GENDER PRONOUNS MATTER?

Pronoun etiquette is in the spotlight. How are pronouns changing? And is it rude to ask a person if they go by he, she, they – or ze?

Sherryn Groch

For words that say so much about us, we probably think about them very little. Gender pronouns, the 'he' or 'she' in a sentence, are almost invisible in everyday English and yet they carry an important piece of our identity.

When actor Elliot Page, of *Juno* and *Umbrella Academy* fame, came out as transgender in 2020, he told the world his pronouns were 'he' or 'they'. Singer Sam Smith in 2019 announced they were non-binary, meaning they do not identify as explicitly male or female – the neutral 'they' is the right pronoun for them.

As our understanding of gender has expanded beyond anatomy (and pink and blue baby clothes), pronoun etiquette has become increasingly front of mind. It's no longer unusual to see pronouns listed beside contact details in email signatures and Twitter bios. Some universities invite new students to share their pronouns when telling the class 'a little about themselves'.

Researcher and writer Quinn Eades recalls explaining his own transition from female to male to his two young children. 'So, you'll still be mama, just a boy mama?' they asked him. 'Exactly,' he said. And he is.

But the new pronoun normal has also sparked a wave of backlash – and lots of confusion – in the public arena. Many are just terrified of getting it all wrong.

So how are pronouns changing? Why do they matter? And how do we use them respectfully?

What pronouns do people use and why?

If you bumped into Melbourne fashion stylist Deni Todorovic at a party, they might say, 'Hi, I'm Deni, my pronouns are they/them/ theirs. Did you watch the footy last night?'

If you wanted to introduce Deni to someone else, you might say, 'Have you met Deni? They LOVE footy!'

It's not a big change, dropping the traditional he or she, but it makes a big difference, Todorovic explains. 'When I came out as non-binary or gender-non-conforming, I didn't change my pronouns right away. I was scared to let go of that last cord of masculinity, I guess. But I'm not completely male or female, I live in the space between so it makes sense to use "they". That feels like me.'

'They' and even gender-neutral pronouns such as 'ze' and 'hir' can be used in place of a male or female designation, for people who 'don't fit into the two boxes on forms', Eades says, and sometimes for those who wish to actively avoid being classified by gender altogether.

It's not unlike the way 'Ms' found its way into the lexicon to avoid women being listed either as married (Mrs) or single (Miss).

At other times, someone 'coming out' as trans, such as Elliot Page, will move to the pronouns that reflect their identity.

When Teddy Cook of LGBTQ health organisation ACON affirmed his own gender as male 15 years ago, he didn't have any language to help him. 'I'm actually a twin but I was the only trans person I knew then – that's so often the case.'

Gender, he says, is largely a presumption made by those around you at birth based on 'what's in your pants', your genitals (known as your sex). 'Most people agree with that presumption, they identify with that gender.'

Usually these people, known as cisgender people, also present to the world in a male or female way, in line with their gender and sex, so they don't declare their pronouns whenever they enter a room. People just tend to guess right.

'But there have always been some of us who don't identify with what we're assigned at birth,' Cook says.

This gender diversity has been a part of the language and culture of Indigenous nations around the world for millennia, he says. 'Think of the Sistergirls [used in many Aboriginal cultures for trans women] and "two-spirit" for non-binary people in Native American [nations].'

Some people may not connect with the transgender term at all or may be living 'stealth', keeping their transition private. Others might use 'rolling pronouns' meaning 'they' might feel as comfortable as 'he'.

> *This gender diversity has been a part of the language and culture of Indigenous nations around the world for millennia.*

Eades, who identifies more as trans masculine than a trans man, goes mostly by 'he' but doesn't mind being called 'they' either – so long as it's not 'sir' or 'buddy' or 'champ'. 'I don't like all those masculine ways men hail each other. I never got called chief or boss before I transitioned.'

Why does it matter what we use?

When pronouns don't match our gender identity, it can be disruptive, even dangerous. The deliberate 'misgendering' of trans or gender-diverse people by, say using the wrong pronoun or by using their original name (deadnaming), has been linked to spikes in mental health concerns, including suicide. This is a vulnerable community already much more likely than the general population to face violence and bullying, to be murdered or suffer from conditions such as depression.

'But when people get [your pronouns] right it's so affirming, it's hard to describe how it feels,' Cook says. 'This isn't about political correctness or snowflakes. Imagine if someone relentlessly called you by the wrong name or gender – even just for one day.'

Because pronouns speak to our identity, they are inherently bound up in emotion, Todorovic adds. 'My parents still struggle to use my pronouns, they don't want to let go of their son. It's really personal, it's about family relationships, too. But that's exactly why we need to get them right. For the same reason, if I know your name is Sherryn, I would feel bad if I called you Sharon.'

> *Because pronouns speak to our identity,*
> *they are inherently bound up in emotion.*

But my English teacher told me 'they' can't be singular?

An email lands in your inbox from Sam. A Sam you don't know. What do they want?

A penguin is born in a zoo. They won't show markers of gender until adulthood. Their name is Ziggy.

Did you notice anything strange about these sentences? Generations of tyrannical English teachers might be screaming that old rule in your ear – that 'they' usually means more than one person. But, as linguist Kate Power explains, the singular they is not actually wrong, however much 'linguistic snobbery' it still provokes. It has appeared in the written language as far back as Chaucer in the fourteenth century, as well as in Shakespeare and Austen.

If you don't know Sam's gender when their email arrives, it's natural to refer to them as 'they'. If Ziggy the penguin has no gender identifiers until adulthood, why pick a pronoun at random?

'"They" fell out of favour as a singular pronoun but it's having a comeback now,' Power says. 'It can indicate multiple people but it can also indicate ambiguity or diversity of gender.'

It might be confusing, she says, but many languages don't have gender pronouns at all. Others, meanwhile, have genders for nearly everything, even objects. In French, for example, a table is female but an office is male.

Besides, language is not just about rules, Power says. It's about being human, too. 'Some people want to politicise pronouns and make them [carry meaning] about [biological sex] instead of gender. That's not what this is about. Whether or not you feel comfortable is irrelevant. The gender a person identifies with is the important bit. That's what pronouns are meant to do.'

In 2019, the singular 'they' was named word of the year by the dictionary Merriam-Webster, and some think it's the only pronoun we need. But Cook stresses most people still identify as one gender or another, including people who have transitioned. 'Being called by the right pronoun after transitioning is really important for a lot of people. It was for me.'

Power agrees 'they' should not be the default, even if it's often the easiest option. 'That's another assumption. This is about listening to each other.'

Will pronouns such as ze and hir catch on?

Cook and Eades think it unlikely but Power says the strength of new pronouns, however strange they might seem at first, is that they also come free from our existing grammatical moulds.

'There's not that same shock to the system of "they" being used in ways we're not as familiar with,' she says. 'Language is changing all the time and we usually try to keep up. The women's movement helped us see that doctors aren't always "he" and nurses aren't always "she". And just look at all the words we've added to our vocabulary like "iso" and "contact tracing".'

Family titles can evolve, too, says Isabel Kenner from Trans Pride Australia. 'I'm a trans woman – for the uninitiated, I was born with male primary sex characteristics I do not associate myself with, but clearly they worked because I have three beautiful children. [They] don't use Mum, I am Dizzy. I let the kids choose it when I transitioned, it's a combo of "Dad" and my name "Izzy". I just use Mum with others to reduce confusion.'

Language is not just about rules.

Eades, meanwhile, is excited to have a 'nibling' – which is a gender-neutral term for a niece or nephew. 'These are our [gender-diverse] young people we have to make room for and care for in our language choices.'

Should I ask people for their pronouns?

It might feel awkward but asking in order to avoid making a mistake is always better than misgendering someone, Cook says. So, we should think of that genderless penguin – and never assume? 'I kind of love that. Yes!'

But isn't asking for pronouns rude? It depends on the intention, Todorovic says. 'If you're not singling someone out and being aggressive about it, then it's not rude at all.'

They suggest this handy trick: offer your own pronouns first. Power says this pre-emptive invitation is actually a common linguistic tactic. 'It's taking the risk on yourself. And it doesn't have to be a big, showy thing. If you're introducing yourself, just add your pronouns. "I'm Kate. You can call me she/her. What do you go by?"'

Todorovic adds: 'The more cisgender people who bring it up or put out their pronouns in bios and signatures, even if they're not a surprise, the more normal it is for the rest of us who need it to be talked about.'

What if I get it wrong?

'Everyone stuffs up, myself included,' Todorovic says. 'We're not used to thinking about pronouns so much. There's a lot of rewiring to do.' If you make a mistake, just acknowledge the correction, apologise and move on. 'It doesn't have to be a big deal. Over-apologising just makes things awkward. We know when it's a genuine mistake and when it's something else.'

If you make a mistake, just acknowledge the correction, apologise and move on.

Cook and Eades agree. 'The amount of times I've had to console someone else for misgendering me or how hard they find pronouns . . . ,' Cook says.

If you really are stuck, he suggests acknowledging that too and spending some time practising or researching. 'You can say, "I'm not getting this but I will." When someone reveals who they are to you, it's a big thing. It's worth practising the words they use.'

Eades recalls a dear friend who was still stuck using his former pronouns a year on from his transition. 'It wasn't feeling okay for me any more, so I finally said something. And he . . . went away and found a little workbook online to practise with and he got it right. That was really special for me. There was love in that. This is about using language with love and kindness.'

THE GENDER BASICS
Some terms, to help you out

Cisgender/cis Someone whose gender corresponds to the sex they have at birth. Cis is Latin for 'on the same side as'.

Gender diverse or **gender queer** A broader term referring to someone who does not fit into one of the two binary genders.

Gender dysphoria A medical condition in which someone's experienced gender is different to their physical sex characteristics. Some people may choose to undergo a medical or social transition to address this.

Gender identity The gender you identify with. This can be different to a **gender role** which refers to social expectations of behaviour and expression associated with being male or female. Most cultures associate wearing a dress with femininity, for example, while someone who is **androgynous** may not express a clear gender in their appearance.

Intersex Someone born with physical sex characteristics that do not fit medical norms for female or male bodies.

LGBTIQ An acronym for lesbian, gay, bisexual, transgender, intersex and queer or questioning.

Misgendering Referring to someone by their former name (**deadnaming**) or former pronoun, or making assumptions about their gender identity based on their appearance.

Non-binary Someone who does not identify explicitly or exclusively with one of the two binary genders. They may identify with both (sometimes called **gender fluid**) or none (i.e **agender**) or wish to throw out the notion of gender altogether (**gender non-conforming**).

Sex Refers to the biological characteristics of males and females, such as genitals.

Sex assignment Refers to the categorisation of a baby as male or female at birth based on their sex.

Sexuality Your sexual orientation such as being sexually attracted to the same sex, the opposite sex or all sexes. Someone who is **asexual**, meanwhile, has no or very low sexual interest in others. Sexuality is different to your gender identity or gender expression.

TERF An acronym for trans-exclusionary radical feminist. It generally refers to a minority of feminists who believe biological sex should take precedence over gender identity, and that the term 'woman' can be fully applied only to those with female reproductive organs.

Trans Someone whose gender does not match the sex they have at birth. For example, someone who identifies as female might also refer to herself as a trans woman. Trans in Latin means across, or 'on the opposite side of'.

4

WHAT DOES THE SUN DO TO YOUR SKIN?

We know we should be 'sun smart'
but what does that mean these days?
And isn't a little bit of sun good for you?

Samantha Selinger-Morris

One person dies about every five hours from skin cancer in Australia – that's about 2000 a year. One is diagnosed with the deadliest form, melanoma, every half-hour. In fact, two out of three Australians can expect to be diagnosed with some form of skin cancer by the age of 70. Many Australians, of course, wear hats and rashies and dutifully cover up their children and themselves in the sun. Yet huge portions of our population remain remarkably unclear – complacent, even – about what damage the sun does to our skin.

Two out of three Australians can expect to be diagnosed with some form of skin cancer by the age of 70.

'You can go to our beaches any day of the week and there's thousands of people who are punishing their bodies, breeding their melanomas,' says Professor Richard Scolyer, co-medical director of Melanoma Institute Australia.

Health professionals are not the only ones to take an interest in people's sun worship. When life insurer TAL surveyed 1000 members and potential customers in 2020, just over one in ten believed they were 'immune' from skin cancer; more than a third said they had never had a skin check; and two-thirds could not name the key signs of skin cancer (see the pages that follow). Half of the 3600 people involved in Cancer Council Australia's 2017 Sun Protection Survey mistakenly believed that sunscreen couldn't be used safely on a daily basis.

Australia hasn't had a national sun safety campaign since the memorable Slip, Slop, Slap! from the 1980s was reprised in 2007, with the added Seek (shade) and Slide (on sunglasses). 'Anyone under 13, they've never been exposed to a national skin cancer prevention campaign,' says Paige Preston, chair of Cancer Council Australia's skin cancer committee.

The most recent key change to messaging? 'UV levels are the only indicator that is accurate for when sun protection is needed,' says Preston. 'People often just look at sunny days and temperature to determine their behaviours. Actually, it's the UV levels, these invisible parts of sunlight that we can't see or feel [that matter].'

So, what is it about 'UV' that we need to know? What does the sun actually do to our skin? And don't we need to catch some rays to top up on vitamin D?

What does the sun do to your skin?

The sun emits radiation that travels to Earth where it is absorbed by the atmosphere – except for the radiation that is not. Some of this radiation is ultraviolet. These rays are a source of both nourishment and damage to humans. Proteins in our skin convert the rays into vitamin D3, essential for bone and heart health, but the rays also burn us.

The culprits are two types of ultraviolet rays. Long-wave ultraviolet A light, or UVA, penetrates deep into skin, causing melasma (greyish-brown patches), wrinkles (by damaging collagen fibres) and leatheriness (again, from depleted collagen). UVA can also lead to skin cancer. Shorter-wave ultraviolet B light, or UVB, penetrates the upper layer of our skin and is primarily responsible for sunburn and most skin cancers.

Of course, both UVA and UVB rays induce tanning, too, by triggering melanocytes, which are cells that produce melanin, a brown pigment. Freckles are tiny spillovers of melanin production, and in most cases are harmless. They need to be checked by a doctor only if they are asymmetrical; have uneven, notched or bumpy borders; contain a variety of colours; have a diameter larger than a pencil eraser; or have evolved (changed in size, shape, colour or elevation). Sunspots are flat, brown spots that develop on areas

UV levels are the only indicator that is accurate for when sun protection is needed.

of your skin that are exposed to the sun. They are also harmless, and only need to be checked if they are black; increasing in size; have an irregular border; an unusual combination of colours (often skin-colour and red); or are bleeding.

Solariums emit both UVA and UVB radiation, which is why they are banned in all states and territories except for the Northern Territory, where there are no commercial tanning businesses.

Tan marks are a graphic illustration of the sun at work, showing the contrast between skin that has been exposed and skin that has been covered, whether the difference is created suddenly and skin is lobster red after a day at the beach, or subtly, almost imperceptibly, by incremental exposure over days, weeks and years.

But here's the rub: there is no such thing as a 'healthy tan'. Tanning is the body's stress response to damaged DNA – the skin's attempt to block the radiation. It is, in effect, the skin saying, 'Make it stop.'

There is no such thing as a 'healthy tan'.

'When you go into the sun, if your skin is exposed – no sunscreen or anything over it – your skin is getting smashed by UVB photons, high-energy photons,' says Professor David Whiteman, deputy director of QIMR Berghofer Medical Research Institute in Brisbane. When fair-skinned people, who have almost no protective melanin, go outside without sun protection, their skin 'gets pulverised', he says. 'Under the microscope, you can see that your skin cells are just smashed with UV damage. It's like a car accident in your skin.

'By the time you've got a sunburn . . . your body is sending out, basically, danger systems, saying, "I'm overwhelmed, I've got dead cells all over the place,"' he says.

Meanwhile, enzymes in our bodies work non-stop to repair the damage that UV photons do to our cells and to replace damaged bits of DNA. Yet, in the face of repeated exposure, they can only do so much.

But how does the sun cause skin cancer?

When a damaged cell divides, the damaged DNA is passed on to daughter cells. Over a person's lifetime, the cells divide many times, passing the mutation to more and more daughter cells. The key issue is *where* the damage occurs in the genetic code. Each cell in our body contains 30,000 genes, each with a different function.

'If [the UVB light] hits a gene that, critically, is involved in that process of cell division, then the cell can just keep dividing repeatedly. That's how cancer starts,' says Whiteman. 'Once the cell loses the ability to switch itself off and just behave, once it loses that regulation, it's just open slather. That's what cancer is.'

Repeated sunburn, which is a radiation burn, might be the leading cause of skin cancer but it is just one of several mechanisms through which cells acquire damage that starts the cancer ball rolling. Repeated, incidental sun exposure is another significant driver. 'What is known is that the more times you go out in the sun, and the more times you get damaged, the higher the probability that you're going to hit a bad gene [that controls cell division],' says Whiteman.

Australia's high number of fair-skinned people, the minimal amount of protective clothing that many of us wear, some regions' close proximity to the equator (where UV levels are highest) and high temperatures (often associated with high UV levels) mean, says Whiteman, that 'the odds are stacked against us [that] eventually, somewhere on your body, you'll get a skin cancer'. This partly accounts for why at least two in three Australians will be diagnosed with skin cancer before 70, according to Cancer Council

Australia. By comparison, about one in five North Americans reportedly develop skin cancer in their lifetime.

'It's purely an accident of history,' says Whiteman of Australia's high skin-cancer rates. 'Indigenous inhabitants are all deeply pigmented, and were highly selected for living in the world's most inhospitable continent. It's a pretty tough environment in Australia [comparatively, for people with fair skin].'

And it doesn't take much to get repeated – and damaging – incidental sun exposure, especially over many decades. 'Every time you go out in the sun ... when you get a sandwich at lunchtime, when you go out to park your car, your skin is getting smashed, particularly from November to February, when the UV index is high, and lots and lots of photons are hitting the Earth's surface,' says Whiteman.

Every time you go out in the sun ... when you get a sandwich at lunchtime, when you go out to park your car, your skin is getting smashed.

How bad can skin cancer be?

Around 98 per cent of skin cancers in Australia are either basal cell carcinoma (BCC) or squamous cell carcinoma (SCC). They both start in the top layer of the skin (the epidermis) but rarely spread (metastasise) to other parts of the body, and can usually be surgically removed. They do carry risks, though. Beyond a small chance of metastasising, they can run deep under the skin, requiring the removal of a large field of tissue that can disfigure a person. When former US *Vogue* model Zacki Murphy was photographed after surgery to remove a basal cell carcinoma from her nose, she wrote, 'My friend titled the photo *Anne Boleyn after Beheading.'* (Other surgeries for BCC and SCC can be more discreet.)

Melanoma, which begins in the pigment-making, or melanocyte, cells in the skin, is overwhelmingly the most dangerous skin cancer. If not caught early, it is far more likely to spread rapidly. As *Science Daily* once put it, 'Unlike other cancers, melanoma is born with its metastatic engines fully revved.'

While there are nearly 770,000 new cases of basal cell carcinoma and squamous cell carcinoma in Australia each year and about 12,000 cases of melanoma, you are more likely to die of melanoma. While 678 people died of non-melanoma skin cancer in 2019, more than double that number – 1415 – died of melanoma, a Cancer Council Australia report shows.

Melanoma can often be cured by being removed; 90 per cent of people with a melanoma less than one millimetre thick are cured after it is removed, particularly if it's detected early. And recent advances in cancer drugs that stimulate the body's immune response (immunotherapy) or that attack cancer cells without harming healthy ones (targeted therapy) are enabling people to survive melanoma for longer.

Melanoma can often be cured by being removed.

'A decade ago, if you had melanoma that had spread to your brain, most people would be dead in six weeks,' says Scolyer. 'Now most people who have had melanoma spread to their brain are alive a year later.'

The five-year survival rate for patients with advanced melanoma has risen, too: 'Ten years ago, it was 5 per cent. Now it's 50 per cent.'

As well as our skin, our eyes can suffer damage from UV rays. In fact, basal cell carcinoma can flourish on the eyelids, says

Whiteman, particularly in the inner corners. Our eyes' lenses can be damaged by sunlight and some melanomas can arise in the retina (although rare, it is believed that sunlight is a likely contributor).

It is possible but rare to contract skin cancer that is unrelated to exposure from the sun. Most notably, some people have a variation of the MC1R gene (a key regulator of pigmentation that creates red hair, freckles and fair skin), which gives them a higher risk of developing skin cancer as a result of having faulty DNA repair mechanisms and insufficient melanin.

While noses are a common locale for skin cancer, it can appear anywhere on your body including – albeit rarely – genitals, inside the eyes, mouth, under fingernails and toenails, on the palms of hands and on the soles of feet.

Is having dark skin a protective factor?

'For people with darkly pigmented skin, their lifetime risk of skin cancer is very, very low; their need for protective steps is lower,' says Whiteman. But – and this is crucial – it is far from zero. Melanoma may be more than 20 times more common in white people than African-Americans, according to the American Cancer Society, but the fate of an individual cannot be predicted by a statistically lower probability of getting skin cancer.

'Remember, probability is probability based over a population, not over an individual, so it says nothing about you or me as individuals,' says Scientia Professor Justin Gooding of the University of New South Wales. 'You may have other genetic factors that mean that [the low probability of skin cancer] doesn't apply to you. You might be one of the ones that are more susceptible [to skin cancer] or less.'

Scolyer, who examines hard-to-diagnose skin biopsies from around the world as a diagnostic oncologist at Royal Prince Alfred Hospital in Sydney, adds this: 'Many people get melanoma who've got a low risk ... We know even people with extremely dark skin get melanoma.'

Jamaican reggae legend Bob Marley died of melanoma when he was 36. It began as a dark spot that appeared under his toenail and spread. This is why the Australian guidelines about sunscreen use pertain to all Australians, regardless of their skin colour.

Don't we need the sun for vitamin D?

Vitamin D plays a crucial role in our health: it helps make sure we have enough calcium in our bones, and calcium plays an important role in heart and brain function. Ninety per cent of the vitamin D in our bodies comes from sunlight on our skin, the rest from dietary nutrients such as dairy products, eggs and fish. UVB rays, the same type of wavelengths that are the primary cause of sunburn, make the active form of vitamin D, called D3, in our bodies when they interact with a skin protein called 7-DHC.

Although an imprecise science, the study of vitamin D suggests that people from some cultures may require less vitamin D than others. 'African-Americans may not need as much vitamin D as their white counterparts to make sure they have healthy bones,' says Professor Rachel Neale, head of the cancer aetiology and prevention research group at QIMR Berghofer Medical Research Institute. 'That evidence is emerging but not completely solid. We don't yet know whether everybody needs exactly the same amount.'

At the same time, Australia lacks hard and fast guidelines as to how much vitamin D people need. Neale is developing a project to create some. Until then, the recommendation in Australia is that we have 50 nanomoles per litre in our blood, which Neale says is a figure taking no chances. By this estimate, she says, 23 per cent of Australians are deficient in vitamin D. By less stringent British standards, which recommend only 25 nanomoles per litre, only 6 per cent of Australians are deficient. (Neale says '25 to 30' nanomoles are 'probably enough'.)

The bottom line? Most Australians get enough vitamin D from incidental sun exposure even while wearing sunscreen. (And 'vitamin D is one part to the story,' says Neale. 'Actually, the best thing you can do to look after your bones is exercise. Weight-bearing exercise is equally, if not more, important than vitamin D.')

So what can you do to protect your body from the sun?

What frustrates cancer specialists is how low skin cancer rates could be – and how many people could avoid disfigurement. Rather than looking for cloudless skies and hot weather as prompts to put on a hat and other forms of sun protection, it's UV levels that we need to be taking our cues from. 'A combination of cloudy, overcast days and the temperature lulls people into a false sense of security,' says the Cancer Council's Preston.

Apply sunscreen every morning when the UV level is three or higher. The UV scale runs from zero (Low) to 11+ (Extreme), and levels differ vastly from state to state, depending on the season. For instance, summer UV levels in Darwin will be around a daily maximum average of 12 and 13, whereas they will be from nine to ten in Melbourne and between seven and eight in Hobart. To monitor UV levels where you are, check the Bureau of Meteorology's website, bom.gov.au, or download the bureau's SunSmart app.

Apply sunscreen every morning
when the UV level is three or higher.

The bureau not only reports the level, and what it means (for example, 8 is Very High), but specifies the time window when sun protection is particularly recommended. As a rule of thumb, avoid sun exposure from 10am to 3pm when UV levels are at their highest. Reapply sunscreen if you get wet and otherwise every two or three hours, regardless of where you are, as cover for incidental sun exposure.

Buy sunglasses that meet strict Australian guidelines, as opposed to sunglasses sold in some outlets, which could come from anywhere, says Whiteman. Sunglasses that comply with the Australian sunglass standard – labelled 'sunglasses' or 'special purpose sunglasses' rather than 'fashion spectacles' – and with the Australian standard AS/NZS 1067 – provide eyes with 'substantial protection' against the sun. They are recommended for both children and adults, particularly the wrap-around type that gives protection from the sides.

Consider using new sun-safety technologies such as UV sensors embedded in smart watches and smart phones, which indicate when the UV factor is high and when it's time to reapply sunscreen. There are already some on the market. A plaster-like UV sticker, created by Gooding and his team at the University of New South Wales, changes from blue to colourless when a certain amount of UV light hits it. The UV light triggers food dye embedded in the paper and degrades it; when it is colourless, it is time to reapply sunscreen or seek shade.

Their initial idea, says Gooding, was to create something that parents could use on their children, after applying sunscreen on them, at the beach. 'You have smart watches that have UV filters [which, for instance, alert you to when the UV factor where you are is over 3] but you're not going to put a smart watch on little Johnnie or little Susie,' he says. '[Parents need] a simple, cheap and easy thing to pop on their kids, so they know when to get them off the beach and into the shade.'

The caveat? These technologies won't protect your skin by themselves. 'It's one thing to have a sensor and watch that works and gives the right information; then we have to understand, does that make people stay out [in the sun] longer [or] shorter?' says Whiteman.

Keep an eye on your skin, too – and let experts do so as well. Book in for regular skin checks and see your doctor if you notice any changes. These include the four main signs of skin cancer: crusty, non-healing sores; small bumps that are red, pale or pearly in colour; spots, freckles or moles that have changed in colour, thickness or shape; and the so-called 'ugly duckling' that looks unlike any other spot on your skin. (The above signs are, primarily, indications for melanoma. Non-melanoma skin cancers are usually pink and/or red, can be itchy, bleed and do not heal.)

'There's no limit [where] you're not at risk of getting melanoma,' says Scolyer. 'If you see something on your skin, get it looked at, please, as soon as you see it. Because if you get it early, you'll be cured. For skin cancer, surveillance is right in our face. It's not, like, inside your body and we can't see it. We have the opportunity to look in the mirror each day [and] use that benefit to pick them up early.'

*

For a full list of signs of skin cancer, visit cancer.org.au, and to assess your risk of melanoma, visit melanomarisk.org.au

5

IS AFGHANISTAN REALLY THE GRAVEYARD OF EMPIRES?

It once lured Western travellers with exotic tribal cultures and breathtaking mountain landscapes, yet Australia and the United States fought their longest wars there. How did Afghanistan become a byword for strife?

Maher Mughrabi

As Donald Trump's aides packed boxes for the president's departure from the White House in January 2021, one of the last men he had appointed to an important government job released a statement. Christopher Miller's mission had been accomplished: the United States had withdrawn all but 2500 troops from Afghanistan (it later turned out that the real number was nearer 3500, as some of the US presence was 'off the books').

Trump's decision prompted Republican Senate Majority leader Mitch McConnell to push back. 'A rapid withdrawal of US forces from Afghanistan now would hurt our allies and delight the people who wish us harm,' McConnell told fellow senators at the time. 'It would be reminiscent of the humiliating American departure from Saigon in 1975.'

Yet, as was the case in Vietnam, the wider American public had become exasperated by an 'endless' or 'forever' war in a distant and alien land. Trump came to office in 2016 pledging to bring home the troops and, while his successor Joe Biden promised to reverse much of Trump's agenda, when it came to Afghanistan all he changed was the deadline, giving it a heavy symbolism: September 11 marked 20 years since the terrorist attack that provoked the US invasion.

The speed with which the Taliban supplanted the Afghan government, entering Kabul with scarcely a shot fired in the middle of August, left US diplomats scrambling to justify their policy. When BBC journalist Yalda Hakim – an Afghan-born Australian – had asked Afghan-born US envoy Zalmay Khalilzad in July what Washington would do if Kabul was overrun, he told her, 'Those are abstract questions.' After the takeover, US Secretary of State Antony Blinken rejected the Saigon comparison, saying: 'We went into Afghanistan 20 years ago with one mission in mind, and that was to deal with the people who attacked us on 9/11, and that mission has been successful.'

This attempt to retrofit history glosses over the huge amounts of money invested by successive US administrations in military training, governance and development in Afghanistan. But it also relies on the fact that US policy always lacked clarity. As Douglas Lute, a three-star general who led the war effort under the Bush and Obama administrations, told government interviewers in 2015, 'We were devoid of a fundamental understanding of Afghanistan ... We didn't have the foggiest notion of what we were undertaking.'

Australia's main military engagement in Afghanistan, as part of Operation Slipper, lasted 13 years and was also the longest in our nation's history. It ended in 2014, in the words of then prime minister Tony Abbott, 'not with victory, not with defeat, but with, we hope, an Afghanistan that is better for our presence here'. That hope has been put to the test in the years since by investigations into the conduct of Australian special forces (which did the bulk of this country's fighting in Afghanistan), with the Brereton report in 2020 finding that 19 current or former special forces soldiers – a small fraction of the serving force – should face criminal investigation over 39 alleged murders of non-combatants. Australia's last troops in Afghanistan – 80 personnel in a support role to the Afghan military known as Operation Highroad – departed in June 2021.

Why has a country that was once a stop on the hippie trail come to be defined by war and terrorism? How did troops of the US-led coalition end up being there for decades? Did the CIA create the Taliban? And what might the future hold for Afghans?

Is Afghanistan really 'the graveyard of empires'?
The one thing everyone knows about Afghanistan is that it cannot be conquered. As this story goes, every army since Alexander the Great's in 330BC has had to leave in defeat. The British, who fought three wars in Afghanistan between 1838 and 1919, did much to

popularise the idea of it as inhospitable and unattainable. (Britain lost the first war but the other two could be called draws.) When Arthur Conan Doyle's Sherlock Holmes meets Dr Watson for the first time, he remarks, 'You have been in Afghanistan, I perceive,' and later Watson recalls seeing 'comrades hacked to pieces at [the 1880 battle of] Maiwand.'

In his poem 'The Young British Soldier', Rudyard Kipling had the last grim word:

When you're wounded and left on Afghanistan's plains,
And the women come out to cut up what remains,
Jest roll to your rifle and blow out your brains
An' go to your Gawd like a soldier.

Three things should be said about this account. Firstly, while travellers and mapmakers had for centuries talked about 'the land of Afghans', or Afghanistan in Persian, there wasn't a state called Afghanistan until the eighteenth century.

Secondly, by that time the area had been conquered repeatedly by Persians, Arabs, Mongols and Indians – peoples whose history is not always closely studied in the West.

Thirdly, and most importantly, Afghanistan is a place on the edge of every place, a landlocked crossroads at the easternmost limits of the Middle East, the westernmost limits of the Far East, the northernmost limits of South Asia and the southernmost limits of Central Asia. Empires haven't been killed there, but they have often found the end of their reach.

This has meant that, again and again, they have sought controlling influence over Afghanistan's affairs without having to rule there, or as Britain's viceroy in India, Lord Curzon, put it in 1906: 'We do not want to occupy it, but we also cannot afford to see it occupied by our foes.'

Map by Jamie Brown

What about the people who live there?

'As an Afghan you are always more than one thing: your kin, your tribe, your ethnicity and the place you were born ... As a foreigner, you can never truly understand what it means to be an Afghan.'

So wrote Abdul Salam Zaeef, one of the founders of the Taliban. At a crossroads, one often encounters a kaleidoscope of people, and Afghanistan has long been home not only to Pashtuns such as Zaeef (once the only people the name 'Afghans' referred to) but also Persian-speaking Tajiks and Hazaras, Turkic-speaking Uzbeks and at least a dozen other ethnic groups.

Its remoteness also gave it allure for Western travellers. In 1955, to promote its new line of go-anywhere jeeps, Land Rover sponsored the Oxford and Cambridge Far Eastern Expedition, which saw graduates travelling overland from Europe along the long-defunct Silk Road. They were supplied with film to record their journey by a young producer at the BBC, David Attenborough.

Programs such as the ones created from their footage and books such as Eric Newby's *A Short Walk in the Hindu Kush* – a mountain range that runs through Afghanistan – inspired waves of young people to set off on a quest to the 'mystic East'. Eventually this would become the 'hippie trail', featuring the hashish Afghans had always grown and smoked.

One visitor particularly awestruck by the country's cities, mountains and wilderness was Tony Wheeler. He and his wife Maureen drove across Afghanistan en route from London to Sydney and wrote a booklet about the journey, *Across Asia on the Cheap*, in 1973, which in effect founded the Lonely Planet publishing empire.

In the 1960s and early 70s, visitors would have found traditional rural societies as well as cities where young Afghans wrestled with modernity in all its forms, from Marxism to the miniskirt. Indeed, it is said that when short-lived national security adviser H.R. McMaster wanted to convince Trump that the country was not irredeemable and that troops should be kept there, he showed him a photo of three miniskirted students on a Kabul street in 1972.

There was one other constant in Afghan life: Islam. For the hippies it would have been part of the exotic furniture. But as the 1970s progressed and conflict and dictatorship reared their heads, for many Afghans it became the last safe space for political activity.

Why did the Americans get interested?

When we look at a map today, Afghanistan is bordered to the north by a number of other Central Asian 'stans. But for most of the twentieth century those republics were part of the Soviet Union, ruled from Moscow. Just as the British in India had kept a nervous

Getty Images/Lawrence BRUN

When former US president Donald Trump argued that Afghanistan was a lost cause, his national security adviser H.R. McMaster is said to have shown him this photograph of miniskirted students in Kabul in 1972 as proof that a different future was possible – if US troops stayed put. Western intervention in Afghanistan and the wider region has often been justified with reference to women's rights. The face-covering burqa enforced by the Taliban became part of the moral case for overthrowing their regime.

eye on the Russian Empire during the nineteenth century in what became known as The Great Game, after the Second World War the United States saw Afghanistan through the lens of its Cold War domino theory, where the fall of one country to communism would mean the same for its neighbours.

By the late 1970s, despite investment by Washington in agriculture, schools and power plants, Moscow had the upper hand, and in 1978 a coup brought communists to power in Kabul. Their efforts to supplant tribal authority and enact land reform provoked resistance. At first Moscow refused to become directly involved but, after a second coup in September 1979, Soviet troops entered Afghanistan to install a loyal government, closing the border and forcing 'hippie' travellers to make a detour.

The Soviet invasion presented Washington with an opportunity to create a Vietnam-like quagmire for Moscow. US president Jimmy Carter had already begun covertly funding anticommunist militants in Afghanistan through the CIA's Operation Cyclone. The invasion prompted the US-led boycott of the Moscow Olympics. Under Carter's successor, Ronald Reagan, Cyclone's budget would grow from hundreds of thousands to hundreds of millions of dollars a year, and those doing the fighting against the Russians would bring a new word into the English lexicon: mujahideen.

Today, when the word jihad (in this context, a war Muslims consider just) has become feared around the world, it is jarring to hear Reagan refer to the mujahideen – the plural of the Arabic word mujahid, or 'one who undertakes jihad' – as 'the true representatives' of 'a nation of heroes'.

Cyclone's funding would be added to by Muhammad Zia-ul-Haq, the dictator of neighbouring Pakistan, and the oil-rich Arab monarchies of the Persian Gulf. By the time the Soviets finally withdrew, in 1989, billions of dollars' worth of weaponry and equipment had flooded into an impoverished nation where unified government was a memory.

Pakistan's north-west frontier, largely populated by Pashtuns with tribal connections across the border, had seen millions of Afghan refugees flee south in the 1980s. Its main city, Peshawar, became the gateway for those seeking to join the fight to the north.

Who are the Taliban? How are they linked to Osama bin Laden?
The rhetoric of the war against the 'godless communists' meant the mujahideen had always included Afghan youths who were originally students of Islam rather than warriors. The Pashto word for such a student is talib, and taliban is plural. When

the communists in Kabul finally fell in 1992, only to be replaced by ethnic and party-based warfare between the mujahideen, a number of these students in 1994 began an armed movement in southern Kandahar province to end internecine strife, making rigid application of Islamic doctrine and courts their hallmark. Swearing allegiance to a religious teacher and mujahid called Mohammed Omar at a tiny whitewashed mud mosque in the village of Sangisar, they took the name Taliban.

For war-weary Afghans, especially in the rural and majority-Pashtun south, the destruction of arms stockpiles in the name of national unity as well as calls for an end to bribery and lawlessness resonated strongly. Within two years the Taliban had swept north into Kabul and declared a new state, the Islamic Emirate of Afghanistan. It would be formally recognised by three countries: Pakistan, Saudi Arabia and the United Arab Emirates.

Another group had been disenchanted by the descent into factional war: Muslims, particularly those from Arab countries, who had formed a kind of international brigade against the Soviets. In 1988, a number of these 'Arab Afghans' became a clandestine Islamist group called al-Qaeda ('the Base').

After the Soviet withdrawal and with foreign funding drying up, many would go home or move to other war zones in the Muslim world. But one would eventually come back: Osama bin Laden, al-Qaeda's Saudi Arabian founder. By the time he and his followers returned in 1996, he was a fugitive from US and Saudi authorities and the Taliban were rulers. In a fateful decision, they agreed to grant him safe haven.

Was the war always unwinnable?

It is unlikely we will ever know for certain how much the Taliban leadership knew about al-Qaeda's horrific plan to use passenger jets as explosive devices targeting the World Trade Center and the Pentagon on 11 September 2001. But by the time the United States and its allies began their 'war on terrorism' that October, both the Taliban and al-Qaeda were in their sights.

It was clear that this would mean an end to the Taliban's 'Islamic Emirate'. But what else it might mean seems to have been unclear to even the most senior political and military planners of the US-led coalition.

There were few places the Taliban could look for sympathy. Their rule had become a byword for religious oppression, excluding women from public life and girls from education, banning traditional and modern pastimes, from kite-flying to television watching, and meting out savage punishments in the name of Islam. The expressions of support Taliban officials extended to Kashmiri Muslims fighting Indian rule, Chechen Muslims fighting Russian rule and Uighur Muslims in Xinjiang fighting Chinese rule alienated those regional powers. The destruction of two giant sixth-century statues of Buddha in Bamiyan set the seal on the Taliban's reputation for barbarism.

But just as invading Afghanistan showed the limits of Washington's reach, overthrowing the Taliban exposed the country's old divisions; the power struggles of the mujahideen leaders returned. At an assembly to ratify a new constitution in 2003, a young woman, Malalai Joya, reminded them of their role in the 1990s bloodshed, saying they might be forgiven by Afghans but not by history. She was ejected from the assembly.

Caught between a series of elected governments dependent on Western support and an insurgency that wouldn't end, US president Barack Obama announced a 'surge' that brought foreign troop numbers in Afghanistan to 130,000. But after the killing of Bin Laden in Pakistan in May 2011, both the political and public appetite for such a large commitment waned. The Taliban kept up attacks on foreign and Afghan forces and a campaign of terror against civilians, using the long Pakistan frontier as a rear base just as the mujahideen had against the Soviets.

The Taliban have insisted all along that they are the legitimate government of Afghanistan. But the seeds of their return to power were sown when US defence secretary Robert Gates revealed in June 2011 that Washington was talking to the insurgent group. By 2013, the US had withdrawn most of its troops and the Taliban had been allowed to open a 'negotiations office' in the Gulf state of Qatar. In February 2020, the US's Zalmay Khalilzad and the Taliban's political leader Abdul Ghani Baradar signed an accord there aimed at finally ending the US invasion.

In that document, Washington had to adopt a contorted diplomatic formula to describe those it was dealing with: 'The Islamic Emirate of Afghanistan which is not recognised by the United States as a state and is known as the Taliban.' But the reality was simple: the Taliban were back in the geopolitical game.

So what could happen now?

It is unlikely that the Taliban's return to power will be a carbon copy of their first rule, but the ways in which it might vary depend on three factors: how much the Taliban have changed in the past two decades; how much Afghan society has changed; and how much the attitudes of the international community have changed.

The Taliban still adhere to their belief that a punitive, puritanical reading of Sunni Islam under the leadership of a 'commander of the faithful' and his select coterie of advisers is the best way to unify Afghanistan, and are still likely to continue to deal with opposition to their rule and policies with draconian corporal punishments. And, despite their commitments to the US, the Taliban retain strong links to al-Qaeda, links that the US contended had allowed al-Qaeda to flourish in Afghanistan in the first place.

Meanwhile, a generation of young Afghans who have, however fitfully, been part of a society promising parliamentary representation, diversity and inclusion of women and minorities will be forced to renounce their gains. The ability of Afghan civil society to oppose the kind of oppression that the Taliban meted out in the 1990s was always dependent on international support. Malala Yousafzai, who was shot as a schoolgirl by the Pakistani version of the Taliban and became a Nobel laureate through her campaign for the rights of women and girls, expressed 'complete shock' at the Taliban takeover: 'I am deeply worried about women, minorities and human rights advocates. Global, regional and local powers must call for an immediate ceasefire, provide urgent humanitarian aid and protect refugees and civilians.'

But her call for a global effort comes at a moment when the international community is deeply divided on questions of democracy and human rights. In the 1990s the Taliban were international pariahs, but Washington's rivals – particularly Moscow and Beijing – may view them differently now. If two permanent members of the UN Security Council were to recognise the Islamic Emirate, where would that leave the West?

After decades of war, Afghanistan has become dependent on billions of dollars in international aid, with more than half the population below the poverty line. Some economists have suggested that a way out of this dependency lies in untapped minerals, with even our own Andrew 'Twiggy' Forrest holding talks on copper, hydropower and geothermal projects with now-ousted president Ashraf Ghani in 2020. Others have argued the whole notion of an Afghan mining boom is a mirage given the country's security situation and lack of infrastructure. Chinese attempts to begin mining copper in Afghanistan had stalled, but in July 2021 the Taliban leadership were received by foreign minister Wang Yi in Beijing. Already a major investor in neighbouring Pakistan, China may see opportunities for its Belt and Road Initiative and to increase its regional influence. Like Pakistan, the Taliban may be prepared to look the other way on China's oppression of its own Muslim population in Xinjiang. Taliban spokesman Suhail

Shaheen told the *South China Morning Post* at the time that 'China is a friendly country and we welcome it for reconstruction and developing Afghanistan ... we care about the oppression of Muslims, be it in Palestine, in Myanmar or in China ... But what we are not going to do is interfere in China's internal affairs.'

Pakistan, which recognised the first Taliban emirate in the 1990s, sees involvement in Afghan affairs as vital to its security. It also worries that cross-border solidarity between Pashtuns might threaten its territorial integrity. If the current Taliban leadership can allay those concerns as well as curbing support for Islamist militancy within Pakistan, Imran Khan's government will not hesitate to deal with them.

Throughout the past two centuries, world powers have interfered in Afghanistan for their own reasons. The Taliban's return and what it chooses to do will force a reassessment of priorities. But it will pose a more fundamental question: what agency do the Afghan people have in setting the course of their nation? In a statement prior to entering Kabul in August, the Taliban said, 'We want all Afghans, from all walks of life, to see themselves in a future Islamic system with a responsible government that serves and is acceptable to all, God willing.' Yet those brave enough to question unity under the Taliban's banner have three options: exile, acquiescence or armed resistance. The decisions that ordinary Afghans make – and the refugee flows that follow – have already put the international community back at the crossroads.

6

HOW CAN PSYCHEDELICS HELP TREAT SUFFERING?

The active ingredient in magic mushrooms is one of the 'trippy' drugs showing promise in treating depression, end-of-life terror and more. What might psychedelics reveal?

Samantha Selinger-Morris

Most of us have an idea – or think we have an idea – of what psychedelics do to us. At their trippy best, drugs such as LSD and magic mushrooms can lead you to feel at one with the universe and awash with creativity. It was under the influence of LSD and peyote derived from cactuses that author Ken Kesey, for instance, wrote *One Flew Over the Cuckoo's Nest*, tipped paint into a stream and dipped T-shirts in it (creating tie-dye), and discovered the world was 'a hole filled with jewellery'.

There are, of course, bad trips, the extremes of which are perhaps best described by the late Boston crime boss James 'Whitey' Bulger who, while incarcerated in an Atlanta penitentiary in 1957, was forcibly injected with LSD as part of the United States' Central Intelligence Agency's now-infamous behaviour control experiments. 'We experienced horrible periods of living nightmares and even blood coming out of the walls,' Bulger wrote.

In 1963, psychology professor Timothy Leary, who had started listing his profession on academic forms as 'ANGEL', was booted out of Harvard University for his research into psychedelics, notably LSD and psilocybin, the active ingredient in magic mushrooms. He gave the psilocybin to undergraduate students when the university had agreed he could only give it to graduate students in his studies – as a tool in psychotherapy and 'mind expansion'. While psychedelics were a hallmark of the counter-cultural hippie movement, by 1970 they had been criminalised, thanks to president Richard Nixon's war on drugs.

In Australia, psychedelics were used by doctors to treat various psychological conditions – none of it systematically documented – but they began to be criminalised from 1970 onwards and remain, mostly, illegal. So, how is it that we're now in the midst of a psychedelic renaissance?

In March 2021, the Australian federal government announced it would provide $15 million for clinical trials to determine whether psilocybin and MDMA could help treat debilitating mental illness. In July, a new privately funded research centre was launched in Melbourne to develop psychedelic medicines. Meanwhile, leading research is underway in a Melbourne hospital into the use of psilocybin to treat end-of-life anxiety and depression in terminally ill patients.

These trials follow an enormous amount of psychedelics research over the past 20 years, in the United States and Europe, which has led to promising findings about the role they might play in treating conditions ranging from severe depression and post-traumatic stress disorder (PTSD) to addiction, Alzheimer's and anorexia nervosa. In fact, in 2021 there are about 100 psychedelics trials worldwide, at prestigious institutions such as Johns Hopkins University in the United States, which launched a Centre for Psychedelic and Consciousness Research in 2020, and Imperial College London, which opened its Imperial Centre for Psychedelic Research in 2019.

What are these studies hoping to find? Could these mind-altering drugs be the long-sought answer to alleviating suffering caused by mental illnesses where other treatments have failed? Do they reveal the secrets of the universe? And what are the risks?

What are psychedelic drugs and what can they do to us?

Psychedelics refer to a broad range of pharmacological compounds that include LSD (D-lysergic acid diethylamide), psilocybin, ayahuasca (pronounced aya-washka, a South American psychoactive brew traditionally used in shamanistic ceremonies), MDMA (known as ecstasy) and ketamine (often called Special K). Broadly speaking, they work by simulating, suppressing or modulating the activity of the various neurotransmitters in the brain – the body's so-called 'chemical messengers' – which carry messages between nerve cells, keep our brains functioning and affect various psychological functions such as fear, pleasure and joy. This leads to a temporary chemical imbalance that results in euphoria and hallucinations, among other things.

The term 'psychedelic' – from the Ancient Greek words *psyche* (soul) and *deloun* (to make visible, to reveal) – was coined in the 1950s by British psychiatrist Humphry Osmond, who was part of a small group of psychiatrists researching the therapeutic potential of LSD in treating alcoholism and other mental disorders.

But even today, researchers don't know exactly how psychedelics do what they do to our brains. 'It's like 10 blind people feeling an elephant from different angles – none of them has the full picture,' says University of New South Wales's Dr Colleen Loo, a professorial fellow at the Black Dog Institute, who is the foremost Australian researcher of the use of ketamine. Ketamine is a powerful animal and human anaesthetic that is legal in Australia, with limitations, and is being used to tackle depression that's resistant to other treatments. One thing researchers know is that ketamine promotes the growth of nerve cells that have shrunk in the brains of people with depression.

> *Researchers don't know exactly how psychedelics do what they do to our brains.*

The fact that the broader picture is far from complete isn't a red flag in itself, given we commonly take medicines even though doctors don't know exactly how they work. 'In a sense, knowing how it [a drug] works becomes probably less important if we know that something works, and if we know that treatment safety profile,' says Dr Vinay Lakra, president elect of the Royal Australian and New Zealand College of Psychiatrists.

For one thing, psychedelics have been shown to increase cognitive flexibility. This doesn't so much refer to their more spectacular effects but rather the way they disrupt our 'default mode network' – the part of the brain that is active when we are at rest, where we think about the future and the past and integrate things that have happened to us. It is central to defining who we are.

In many people, the stories they tell themselves are embedded in rigid and destructive patterns of thought: people with depression or anxiety often tell themselves they're worthless and unlovable. Those who have experienced trauma often feel a sense of survivor's guilt.

Psychedelics, in conjunction with psychotherapy, temporarily short-circuit these ruminative, and frequently negative, mental loops. In doing so, they help provide new perspectives on old problems. Michael Pollan, a Berkeley journalism professor who chronicled his encounters with psychedelics in a 2018 book *How to Change Your Mind*, spoke to one man who had quit smoking after a psychedelic trip 'because I found it irrelevant'.

Many trippers also describe a temporary breakdown of identity, erasing the distinction between self and non-self, which also seems to have benefits. As people involved in a psilocybin smoking-cessation trial variously recounted to Pollan, 'I died three times. I sprouted wings. I flew through European histories. I beheld all these wonders. I saw my body on a funeral pyre on the Ganges. And I realised, the universe is so amazing and there's so much to do in it that killing myself seemed really stupid.'

Other people in clinical psilocybin trials report radically heightened senses, says Dr Margaret Ross, chief principal investigator of a psilocybin-assisted psychotherapy study at St Vincent's Hospital in Melbourne. 'Music can sound amazing, or people can see music,' she says. 'Their visual field is often impacted; they see all the colours, people often see patterns and geometry, or weird sort of things. They can also experience their body very differently, they might feel like they're warping, melting or dissolving.'

'[They're] kind of at one with the universe, if you like,' says Dr Paul Liknaitzky, a research fellow at Monash University who is the lead investigator on four psychedelics trials, 'and exist beyond space and time and thought, and these are often called mystical experiences. It's a little like what astronauts report from looking at the Earth from outer space – this enormous perspective on life that allows people to no longer fear death, or no longer fear anything, because they've got a different perspective on things.'

This is why, says Liknaitzky, psychotherapy-assisted psychedelics are being trialled to treat so many different problems, from end-of-life anxiety and terror to nicotine and cocaine addiction, to PTSD. The drugs address 'more fundamental aspects of psychological distress than just the symptoms' – although it's no reductive pill-popping exercise. The clinical trials almost always include multiple psychotherapy sessions to help address fears and personal history, and to help the patient make sense of their psychedelic experiences. No current trials or experiments simply use the drugs by themselves; in many, the therapy incorporates lifestyle changes such as better eating and exercise.

One woman who participated in a psilocybin trial to combat nicotine addiction at Johns Hopkins – but who ended up confronting long-held grief about the breakdown of her marriage and a miscarriage – told Dr Albert Garcia-Romeu at Johns Hopkins, 'It really helped me move forward.'

'It was almost like a letting go of some grief that had been in there, that she may not have wanted to deal with, or acknowledge,' says Garcia-Romeu, who is involved with numerous psychedelics trials.

Another woman with terminal colon cancer reported that taking psilocybin enabled her to enjoy her final months of life when she was otherwise paralysed from making even the most mundane plans, so mired was she in fear. 'I felt this lump of emotions welling up ... almost like an entity,' she told researchers in the Harbor-UCLA Medical Center trial. 'I started to cry. Everything was concentrated and came welling up and then it started to dissipate and I started to look at it differently. I began to realise that all of this negative fear and guilt was such a hindrance ... to making the most of and enjoying the healthy time that I'm having.'

The same, it's been reported, can go for LSD. *The New York Times* reported on a trial using LSD-assisted talk therapy for people with end-of-life anxiety: 'One 67-year old man said he met his long-dead, estranged father somewhere out in the cosmos, nodding in approval.'

It's not all sunshine, rainbow-enveloped self-acceptance and tripping across the cosmos, though. The work can be confronting, as with much therapy. One woman who was sexually assaulted by her father as a child, and who was suffering from PTSD as a result, ended up 'transferring' her anger and sadness on to a male psychotherapist who was guiding her through an MDMA trial in the United States. 'She was actually quite distressed the next day when she realised what had happened [during the trip],' says Dr Stephen Bright, senior lecturer of addiction at Edith Cowan University, who witnessed the session. But, despite her embarrassment at lashing out, the treatment lessened her PTSD symptoms. (PTSD symptoms include anguish-inducing flashbacks and hypervigilance.)

It's not all sunshine, rainbow-enveloped self-acceptance and tripping across the cosmos, though.

▌ Why the interest in psychedelics now?

'All of a sudden, it hasn't got that kind of "counter-culture" tag,' says St Vincent's Ross. She points to testimonials from high-profile, respected journalists and writers such as Pollan, who was in his 60s when he wrote his book about trying psychedelics. 'I felt my sense of self scatter to the wind,' Pollan told National Public Radio, 'almost as if a pile of Post-Its had been released to the wind – but I felt fine with it. Then I looked out and saw myself spread over the landscape like a coat of paint or butter.' Pollan says the drugs act upon 'the self that talks to the self'.

The year before, author Ayelet Waldman published a memoir about how microdosing on LSD saved her marriage, by, among other things, easing her depression and bipolar disorder. 'It changes the profile of who would use something like this,' says Ross, whose study focuses on whether psilocybin alleviates anxiety and depression in terminally ill patients.

Brain sciences and scans that reveal brain activity have also been trending over the past decade, lending legitimacy to psychedelics studies, says Garcia-Romeu. 'If you can measure something in the brain then you can point to a mechanism, and so understanding the biological and brain-based mechanism is an important part of validating this work from a kind of more hard science perspective, if you will,' he says. (Some researchers now refer to their trials as 'anti-Leary' as Leary's studies were largely anecdotal.)

This modern research dovetails with a growing dissatisfaction within the scientific and medical communities with current treatments for depression and anxiety. Although antidepressants work for many people, they don't for large portions of the population. And they 'really don't offer much' for people with terminal illnesses experiencing end-of-life anxiety and depression, says psychiatrist Justin Dwyer, who is leading the St Vincent's Hospital psilocybin trial, along with Ross. 'For giving people this sense of reconnection, of oneness with family, of meaning, of purpose, of feeling as though there's much more to them than their illness, antidepressants do nothing.' For one thing, he says, antidepressants often take weeks to work, and frequently cause nausea, which might interact with pain medication, and have a sedating effect when people are most longing for connection.

What have psychedelics studies found?

A ground-breaking study from Johns Hopkins in 2016 found that of 51 cancer patients suffering psychosocial distress, 80 per cent who received psilocybin (with psychological support) had significant reductions in depressed mood and anxiety. Data suggests the treatment lowers anguish for as long as six to nine months afterwards, says Garcia-Romeu. A 2020 study from Johns Hopkins showed that psilocybin, taken by people with a major depressive disorder, was four times as effective as standard antidepressants. And another Johns Hopkins psilocybin study, from 2014, reported that of 15 participants, 80 per cent abstained from smoking, over six months. A year later, 67 per cent of participants still abstained. (This compares to a 35 per cent success rate for patients taking Varenicline, a prescription medication widely considered to be the most effective smoking-cessation drug.)

As for treating PTSD, a study from the non-profit Multidisciplinary Association for Psychedelic Studies (MAPS) in California found that, of 107 participants with chronic, treatment-resistant PTSD (who suffered, on average, for nearly 18 years), 61 per cent no longer qualified for PTSD after three sessions of MDMA-assisted therapy. At the 12-month follow-up, 68 per cent no longer had PTSD.

And what of LSD? Twelve people who participated in a Swiss trial of LSD-complimented talk therapy as they neared the end of their lives reported positive results. 'Their anxiety went down and stayed down,' psychiatrist Dr Peter Gasser has said.

And that microdosing that Waldman found so life-changing? There are few studies of microdosing but one of the first, from Macquarie University in New South Wales in 2019, found mixed results. Ninety-eight participants who took 'super low doses' of various psychedelics – including psilocybin, LSD and mescaline – experienced an immediate boost in mood and a feeling of connection with their surroundings as well as significant drops in depression and stress – but no drop in anxiety. They also reported a significant increase in neuroticism – a tendency to feel sad, angry, anxious, or vulnerable – although a subsequent study by Macquarie found its microdosers experienced a decrease in neuroticism.

Almost none of these clinical trials has yet moved to phase three, which typically lasts between one and three years and confirms safety and effectiveness on large populations, comparing the drugs to standard therapies. MAPS published the results of its phase-three trial of MDMA-assisted psychotherapy to treat severe PTSD in 2021. Ninety people participated in the trial: two months after treatment, 67 per cent in the MDMA group no longer qualified for a diagnosis of PTSD compared with 32 per cent in the placebo group.

Professor Michael Farrell, director of the National Drug and Alcohol Research Centre at the University of New South Wales, is glad that research is being done on psychedelics but cites the lack of large-scale trials so far as one reason to be cautious. Small trials make it difficult to prove cause and effect, for example. 'When people say that there's now very strong evidence [for psychedelics' benefits], I wouldn't agree with that.' Wayne Hall, an emeritus professor at the National Centre for Youth Substance Use Research at the University of Queensland, agrees, saying that any new drug can have a 'placebo effect' among participants and practitioners who are 'really convinced of the value of what they're giving'. But he says, 'It is pretty promising that they've [researchers] managed to get clinically significant effects in the trials that they have with numbers as small as they have – that's encouraging.' Larger trials will give a clearer picture of who will likely benefit from psychedelics, and under what circumstances.

What are the risks of taking these drugs?

Psychedelics might not generally, as many experts put it, 'follow the cycle of abuse' – they don't typically lead to addiction, cravings or withdrawal. Apart from anything, unlike addictive drugs such as heroin, which triggers euphoria (at least at first) or benzodiazepines, which melt anxiety, psychedelics have unpredictable and often frightening effects, says Bright, which means people rarely take them as often as one would need to in order to become addicted.

That said, the risks are myriad – for the vast majority of cases, in non-clinical settings – and taking psychedelics can have tragic results. In rare instances, some psychedelics can evoke a lasting psychotic reaction, more often in people with a family history of psychosis. (Participants in clinical trials, who must be above the age of 18, are screened for a family history of psychosis, schizophrenia and other conditions including bipolar disorder and complex trauma, which could lead to damaging outcomes.)

MDMA, which is a stimulant – the MA is for methamphetamine – elevates heart rate, blood pressure and body temperature. 'So, if you're in a panic state, that would increase your panic state,' says Dr Monica Barrett, a social scientist at the University of New South Wales and the lead Australian researcher for the Global Drug Survey, a London-based research company that monitors drug use. In rare cases, MDMA can lead to serotonin syndrome, which causes overheating and can kill, as has been seen at music festivals. And, of course, when people take psychedelics in a non-clinical setting, the hallucinations can lead them to take potentially fatal risks such as walking into traffic or jumping from a high place in the belief that they can fly.

Then there's the reality that many psychedelics being offered by a burgeoning number of sometimes untrained underground psychotherapy practitioners – outside of clinical trials – are not using pharmaceutical-grade drugs. 'That's Russian roulette,' says Bright. 'Their PTSD could actually become worse because they're being re-traumatised rather than reprocessing their trauma. [And] it creates a lot of suggestibility, so you could brainwash people.'

Bright remembers one woman who presented at a Victorian clinic he worked at who had been sexually assaulted two years prior to showing up but had no PTSD before she used psilocybin one day at home with her partner. 'She freaked out, her partner called a paramedic, she was taken to the hospital, the psychedelics were reversed with an anti-psychotic – and she developed PTSD and a drinking problem.'

Recreational users who take high doses of ketamine over long periods report a condition called 'ketamine bladder': the organ becomes so inflamed the lining dies off. Sometimes, the entire bladder needs to be removed. Ketamine also features a risk even in a clinical setting – of relapse. (Dr Loo's trials have not featured a psychotherapy component.) 'He said he'd had this amazing experience,' says Dr Loo, recalling a male participant in the first Australian clinical trial of ketamine for depression. Within four hours of taking the drug, he had started changing before her eyes, as many who take ketamine do: his face was brighter, he became more reactive. 'They've got a spring in their step,' she says. But by the end of the week, the effects had worn off. The antidepressant qualities of ketamine often fade after a few days, and treatment, due to governmental restrictions, is allowed for only two months at a time. 'He said something to me I've never forgotten, that the devastation of the relapse was bigger than the elation of getting well,' says Loo.

For other people, she says, ketamine provides welcome temporary relief as a treatment of last resort when all else, including antidepressants and psychiatric therapy, have failed. 'It's not an easy treatment to manage,' she says, but she's been 'completely astounded' by its impact on some patients. 'It's truly amazing ... You could see the same person yesterday, and today [after treatment] a completely different person.' She next wants to investigate the use of ketamine, this time combined with psychotherapy, to manage treatment-resistant anxiety.

What's next for psychedelics in Australia?

In July 2021, a $40-million global institute was launched to develop new psychedelic medicines and psychotherapeutic treatment models to go with them. The Psychae Institute, in Melbourne, is funded by a North American biotechnology company and aims to connect leading researchers, including those from Swinburne University, the University of Melbourne and the Florey Institute of Neuroscience and Mental Health, to others from around the globe.

Australian experts say that, after years of trailing in the psychedelics research race, they are swiftly catching up. In 2021, six clinical trials of MDMA or psilocybin were planned or underway, including one at St Vincent's Hospital in Melbourne, which is testing psilocybin for treatment-resistant depression and end-of-life anxiety and depression. This study would expand on previous research, as it's being conducted with terminally ill patients who have a number of diseases and conditions, whereas the 2016 Johns Hopkins trial, its most important predecessor, treated only patients with cancer. If a trial eventually leads to treatment approval, the approval is only for the illnesses that it's been trialled on.

Ross, who has long worked in palliative care as a clinical psychologist, hopes the St Vincent's Melbourne study confirms the Johns Hopkins findings because 'people are not coming along to us saying, "I'm depressed and anxious [at end of life]". What they're doing is saying, "I'm terrified. I've lost all sense of meaning and purpose in my life. I feel completely unmoored from everything that gives me a feeling of identity." They may have limited time left and they're spending it anguished and pulling away from people.'

Early in 2021, Australia's drug regulator, the Therapeutic Goods Administration (TGA), rejected a push by the not-for-profit organisation Mind Medicine Australia to allow psychiatrists to prescribe MDMA and psilocybin because of lack of evidence as to its efficacy. It was an interim decision, and the TGA has since sought independent review of the evidence. Many local researchers were relieved. 'There are a lot of evangelistic claims being made but this is no silver bullet,' says Bright about psychedelics, adding that the hype could encourage people to embark on unsafe experiments in non-clinical settings. 'Desperate people will do desperate things – they will go, "How will I find this?"'

However, Bright adds that, historically, the TGA has followed rulings by the US Food and Drug Administration and, increasingly, the FDA is supporting psychedelics. In 2017, it granted 'breakthrough therapy status' to MDMA-assisted psychotherapy, meaning it will develop and review it faster than other candidate therapies for PTSD. Bright predicts the FDA will approve MDMA-assisted psychotherapy for PTSD in the States by 2024, with Australia following suit after that. Garcia-Romeu predicts that, in the United States, laws allowing psilocybin-psychotherapy assisted treatment for depression 'will probably be five years [away], maybe'.

In the meantime, parts of the United States are embracing psychedelics. In 2019, the city of Denver decriminalised magic mushrooms, or psilocybin, after a referendum. In 2020, Oregon became the first US state to vote to legalise psilocybin for therapeutic use.

Garcia-Romeu is wary about people believing that psychedelics might offer a 'magic pill that will solve all their problems'. 'I always say, "This is not going to pay your bills, it's not going to wash your dishes, or fix your relationship with your estranged family members."'

'You don't just take a pill, and all of a sudden you're in fantastic shape.'

There are a lot of evangelistic claims being made but this is no silver bullet.

7

HOW DO LIBERAL AND LABOR FACTIONS WORK?

The factions (and fractions) involved in politics in Canberra can be bamboozling. What role do they play? And who's who in each of the major parties?

James Massola

In Roman times, the first triumvirate of leaders – Julius Caesar, Pompey and Marcus Crassus – was, essentially, a faction of three powerful men who worked together to take charge of the Republic and carve up the empire among themselves. When Crassus died in battle in 53BC, Pompey and Caesar turned on each other and Caesar and his faction emerged victorious to rule Rome. He lasted until 44BC, when a 'faction' of more than 60 senators led by Cassius and Brutus murdered him.

Thankfully, in Australia, MPs don't tend to engage in actual assassinations of their opponents – but factions do take down leaders (and in the past decade, frequently and brutally). Yet, despite the fact that we all know there are factions, they are not easy to map: you won't find MPs' factional memberships spelled out on their email signature. To get to grips with which MPs align with which groups, we asked them ourselves, conducting interviews in 2021 with 39 Liberal MPs in the 91-member federal parliamentary Liberal Party and with 25 of the 94 MPs in the Australian Labor Party. Their allegiances are often opaque to the opposing side, so there was a lot of anticipation among MPs keen to see them comprehensively laid out.

Thankfully, in Australia, MPs don't tend to engage in actual assassinations of their opponents.

The response to the publication of the resulting factions maps (see p. 109) was immediate – readers clearly had an appetite for lifting the veil on the 'faceless men' (and women) who run our political parties in the hothouse of federal parliament. Dozens of MPs from all sides texted us to argue that their influence had been under-represented, or that their enemies' influence had been overstated. They also said thanks – more than one noted they had always wondered which faction this or that MP from the opposite benches ran with, and now they knew. An MP from the National Party even jokingly asked if we could write an explainer on the 21 factions inside the 21-member Nationals party room.

So, what did we find out?

What are factions (and fractions) and why do they matter?

The word 'faction', from the Latin 'to do' or 'to make', was used in ancient Rome – to talk about chariot racing. The rival teams whose drivers careered around the Circus Maximus while tens of thousands of spectators cheered were called factions. Today, factions are a common feature of parliamentary democracies, from the United Kingdom to New Zealand and Malaysia.

In Canberra, the factional systems in the Labor and the Liberal parties differ markedly at every level – state, national and among the MPs. The Liberal Party has groupings of interests around policy positions and personalities, which, broadly speaking, coalesce around the Moderates, the Centre Right or Morrison Club and the National Right. All but a handful of Liberal MPs are in one group or another. Labor's factions are more formally organised between a Left and a Right. In 2021, just two federal Labor MPs are not members of a faction.

Both parties have sub-factions (sometimes called fractions) and it's in these splinter groupings that the complexities of political allegiances become really apparent.

The Victorian Industrial Left in Labor, for example, is a small group in the federal parliamentary Labor Party (and the Victorian state parliament) that is actually distinct from the alliance of all state-level Left factions collectively known as the National Left. But while its numbers are low in federal parliament, the Industrial Left retains greater influence through the union movement – for example, the group has a significant bloc of votes it can use at state and national party conferences to help determine policy decisions.

Meanwhile, on the Liberal side in Canberra, there's a 'powerful individual' sub-faction within the broader Morrison Club or Centre Right – an 'ambition faction' of MPs from Victoria close to Treasurer Josh Frydenberg. Frydenberg is whip-smart, an indefatigable networker and is sometimes jokingly referred to as the government's press secretary, such is his willingness to text or WhatsApp journalists and editors from early in the day to late in the evening.

Sometimes, there isn't a lot of difference between two groupings. The Liberals' Centre Right and National Right have much in common philosophically, particularly on taking a conservative approach to social issues, but the differences are a question of emphasis. The Centre Right group describes itself as more pragmatic and outcomes-focused whereas at least some members of the National Right are more likely to fight harder – and publicly – on issues such as opposing marriage equality, abortion access rights or going into bat for religious freedoms.

Within Labor, generally the distinctions are a little clearer: the Left has historically been more progressive, focused on social issues and more supportive of intervening in the private sector, whereas the Right has been more economically dry, more supportive of Australia's alliance with the United States and, in some cases, more socially conservative.

In both parties, factions play a key role in deciding leaders (more on that later). Labor's factions also play a key role in choosing candidates for safe and newly created seats. The Left, for example, will run a more progressive candidate in the Greens-held seat of Melbourne, whereas the Right might stand a candidate in an outer-suburban seat in Sydney that Labor wants to win back from the Liberal Party. And factions help to decide – or, if you will, fight among themselves to decide – the order of candidates on the Senate ballot at an election, where the difference between being first, second and third position can mean a place in Canberra or not. For the Liberals, factions or groupings also play a part in helping choose candidates for seats – the party's rank-and-file members get to vote for a preferred candidate, but those rank-and-file members have allegiances to one or other grouping.

In both parties, factions play a key role in deciding leaders.

For both the Liberals and Labor – although more so for Labor – the factional systems bring order to the party room and, in Labor's case, the caucus (Labor's parliamentary party). In Labor, factions gather like-minded rank-and-file members and MPs into cohesive groups to advance policy positions – for example, to argue for more ambitious climate change policies or for a change in refugee policy. Factional conveners, elected by faction members, manage the egos and ambitions of MPs seeking entry to the ministry, and help parcel out promotions on the basis of merit and quotas (also discussed later). And factions manage internal policy fights, from those over the privatisation of Qantas and the Commonwealth Bank in the 1980s and 90s through to marriage equality and refugee policy in the 2000s.

In the Liberal Party, factions don't play as great a role in giving out ministerial positions – that's very much the gift of the leader of the day – but they are involved in managing policy fights. Again, a good example is the debate over legalising marriage equality, for which the moderates pushed harder and harder and which the National Right fought against tooth and nail.

In both major parties, few women are considered powerbrokers. In part, this reflects the make-up of the parliament itself – about 35 per cent of MPs are women, although Labor (47 per cent women) has far greater representation by women than the Liberals (23 per cent women).

How do Labor factions work?

Labor people have been grouped as left or right for decades, and the factional system began to be formalised in the 1970s. Today, by the time they've entered parliament, MPs will most often have been in a faction for some years. Sometimes, they will have been invited to join a faction along the way, sometimes they might request to join one – the process remains opaque.

> *Labor people have been grouped*
> *as left or right for decades.*

While the majority of Labor's rank-and-file members are Left-leaning, and tend to be drawn to the more progressive wing of the party, the Right usually has more MPs (although the number of Left MPs has grown in the past decade). Of the 94 Labor MPs in our 46th parliament, 49 belong to the Right, 43 are in the Left and two are unaligned. Who receives a frontbench place comes down to the iron laws of arithmetic. Here's how: the Left is currently allocated 14 of the 30 seats in the shadow ministry on a proportional basis that reflects the number of MPs they have – so the Right receives 16. But the two groups give out those spots differently. The Left chooses its 14 frontbenchers from a national list of all its MPs, whereas the Right hands out frontbench spots using a state-based quota formula. That means the New South Wales Right, for example, gets six spots in the shadow cabinet (it has the highest number of MPs), the Victorian Right gets four, the Queensland Right receives two places and Western Australia, South Australia, Tasmania, the Australian Capital Territory and the Northern Territory are treated as one bloc – and take the final four frontbench spots.

But it's in choosing a leader that the factions come into their own. A Labor MP cannot rise to the top spot without factional support. Former prime ministers Gough Whitlam (1972–75) and Paul Keating (1991–96) were both from the NSW Right, which considers itself the king-making faction of the ALP. Whitlam's factional alignment with the NSW Right was important, but the factional system was not as formalised then as it is now. Keating, on the other hand, was a card-carrying member of the faction at a time when it was arguably at its most influential, with people such as senator Graham 'Whatever It Takes' Richardson working in the background to support him. Former PM Bob Hawke (1983–91) was in the centrist Victorian Right but was installed with the support of the NSW Right, too.

Victorian former prime minister Julia Gillard (2010–13), while notionally in the National Left, was actually part of a predominantly NSW sub-grouping called the Soft Left or Ferguson Left, named for former MP Laurie Ferguson. Like the Industrial Left in Victoria, this sub-grouping – which now includes only a handful of federal MPs – has historically done deals with the Right to secure seats and ministerial posts. It's this alignment that helps explain why large sections of the NSW, Victorian and Queensland rights backed Gillard to replace Kevin Rudd in 2010. (Rudd was aligned with the Queensland Right – specifically, a small sub-group known as the Old Guard. This was the smaller of two Queensland Right sub-factions at the time – the larger grouping was the AWU Right, led by unionist Bill Ludwig for decades, which is still the dominant group in the Sunshine State.)

Fast-forward three years and Gillard's removal and Rudd's reinstatement was made possible by a significant group in the Right deserting Gillard. After retaking the leadership, Rudd brought in sweeping changes to democratise the ALP and reduce the power of factions, including giving members outside the parliamentary party a 50 per cent say in the leadership vote and making it harder for the caucus to replace the leader. So, in theory, the topplings we saw in 2010 and 2013 can't happen again – but that's yet to be tested. In 2020, Rudd went a step further and called for the abolition, even banning, of factions.

> *So, in theory, the topplings we saw*
> *in 2010 and 2013 can't happen again.*

How do Liberal factions work?

Former Howard government minister David Kemp wrote the 1973 essay 'A Leader and a Philosophy' and is a good person to explain how the Liberal Party works, and why the Morrison Club has grown.

'Liberal factions tend to be very different to Labor's factions,' Kemp says. 'They aren't organised in the same way and they tend to cut across each other on different issues, it's not as clear-cut and there is much more fluidity. The groupings in the Liberal Party tend to be around personalities and issues, but you get people with similar views in different groups.'

In the 1980s and 90s, the dries emphasised free-market economics and conservative social policy while the wets favoured more progressive social policy and bigger government. Those two groups are no more. These days the Moderate faction, which meets semi-regularly during sitting weeks, is the leading advocate of free-market economics, whereas the National Right (or Hard Right), which meets regularly, is more concerned with social issues: religious freedoms, gender identity, national security and, until recently, opposing marriage equality. The National Right is the most organised faction in the Liberal Party and, as one of its members puts it, it believes in government that is as big as it has to be and that the future of the Liberal Party is in the outer suburbs and the regions.

On a state-by-state basis, the Liberal factional system in NSW and South Australia is relatively formalised. NSW MPs are divided among the three groups, and South Australia has Moderates and National Right members. These two states are the most similar to the ALP in terms of structure. Across the rest of the country it's more complex. MPs in Victoria tend to gather around powerful individuals – Frydenberg and, in the past, former treasurer Peter Costello. The running joke among party Moderates is that in Queensland, WA and Tasmania there are no Moderates – like Canberra's version of the Tasmanian tiger, there are people who swear they exist but they rarely break cover in public.

The Centre Right or Morrison Club doesn't meet regularly. It is really several overlapping groups holding shared interests with Morrison as its figurehead. It's about proximity to power, loyalty to the leader, shared values and a social outlook. The group's unifying philosophy is pragmatism – an adherence to free-market economics but with enough flexibility to splash billions to prop up the economy during the COVID-19 pandemic – and relatively conservative social values. As one member of the group puts it, 'We realise you don't win elections by yelling at people about abortion. We are dry economically and socially conservative but not in an "in your face" way.' Philosophically, the National Right and the Morrison Club have a lot in common – just different emphases on specific policies – and many members could be at home in either group.

Climate change is still a lightning rod for the Liberal Party: the Moderates favour stronger action to mitigate it, the National Right counts some sceptics among its number, and the Centre Right takes a pragmatic, middle-of-the-pack approach (as it does on many other issues).

And Liberal factions also surface during leadership contests and challenges, when MPs have to nail their colours to the mast. Scott Morrison's ascent was assured by his tight inner circle (before the Morrison Club ballooned in size) working with the Moderates, who backed him rather than their own candidate, former deputy leader and foreign minister Julie Bishop, to ensure Peter Dutton (from the National Right) did not become prime minister. In ordinary times, the huge spending Morrison's government unleashed in response to the pandemic would have caused a civil war in the Liberal Party, but Morrison has managed to present the big spending as pragmatic and necessary during such an extraordinary and challenging event.

Who are the powerbrokers in the Liberal Party?

In the Liberal constellation of alliances, personalities accrue loyalty over time. Morrison's inner circle primarily consists of MPs who entered parliament when he did, in 2007: Alex Hawke, the factional organiser of the Centre Right in New South Wales, Queenslander Stuart Robert and West Australian Steve Irons. This core group is also defined by their shared faith (all are members of the Prayer Group – more on that later) and their 'let's get things done' approach. Ben Morton, another West Australian and, like Morrison, a former state division director, is one of the PM's closest and most able lieutenants, but if Morrison were not in the parliament, Morton's philosophical home would be the National Right.

About half of Treasurer Josh Frydenberg's Victorian group – including Greg Hunt, Dan Tehan and a handful of others – belong to the Morrison Club. So do a large cohort of Queenslanders, many of them first-termers, people of faith or both, and some MPs from other states. Both Moderates and National Right members argue that some MPs self-nominate as members of the Centre Right because it's the Morrison Club – personal loyalty to the PM matters in winning favour and potentially promotion. Much as the 'class of 1996' – the Liberals elected in 1996 – were loyal to John Howard, so too are the 15 new MPs elected in the 'class of 2019' loyal to Morrison for winning the unwinnable election.

Nowadays, the PM has broad support from all three factions. But that doesn't mean the Moderates are always happy with his policy approach. Some feel he takes the Moderates' support for granted; and they believe that, in a post-Morrison era, 'a lot of people in the Centre Right, in particular, will move back to us'. On the day Morrison leaves parliament, it is likely the Centre Right will begin to lose members back to the other two factions unless Frydenberg, the clear heir apparent to the Liberal leadership, can hold the club together. Back in 2014, there was a (Joe) Hockey Club – but political trajectories can reverse in an instant.

Meanwhile, the three most influential Moderates in the federal Liberal Party are Simon Birmingham, Marise Payne and Paul Fletcher – all cut from similar cloth but with quieter personalities than their factional predecessors Christopher Pyne, Julie Bishop and George Brandis. Within the Moderates, the New Guard or Modern Liberals is made up of MPs elected in 2016 or 2019, who typically represent inner-city lower house seats or are in the Senate, and who are more economically dry than their older colleagues, progressive on social issues and, if anything, willing to advocate for more ambitious climate change policy. As one member puts it: 'We are Menzies Liberals, the "live and let live" people. And we are the old dries and wets at the same time.' The New Guard has landed the chairmanships of some of the 46th parliament's most important committees, including mental health (Fiona Martin), taxation (Jason Falinski), economics (Tim Wilson), fintech (Andrew Bragg) and treaties (Dave Sharma).

The National Right is undergoing a period of change. For years its figureheads were Abbott, Tasmanian senator Eric Abetz and former defence minister Kevin Andrews. Abbott left the building, Andrews' long stint as an MP will end with the 46th parliament because he lost preselection for the safe seat of Menzies in Melbourne, while Abetz has entered the twilight of his Senate career.

Peter Dutton is the notional leader of the National Right – and is perhaps the least well-understood conservative in the parliament. He is a 'national security conservative' rather than a religious conservative and is, arguably, more socially progressive than the Prime Minister. It was Dutton, for instance, who helped engineer the marriage equality postal survey in 2017. The wily Queenslander personally opposed the law change but he could see the issue had to be dealt with to take it off the agenda as internal ructions were strangling the Turnbull government. Underscoring this pragmatism, Dutton voted no in the survey but yes in the parliament.

Mathias Cormann and Dutton – close friends and allies – led the National Right for years and Cormann's decision to quit in 2020 left a big gap in the group. Victoria's Michael Sukkar and the ACT's Zed Seselja have since grown in influence in the faction, while Angus Taylor, as a member of Cabinet, has seniority but doesn't tend to wield influence in factional brawls over preselections.

Who are the powerbrokers in the Labor Party?

While Labor factional conveners do some of the grunt work of organising and running meetings, they are not necessarily the powerbrokers. The member for Kingsford Smith, Matt Thistlethwaite, is the national convener of the Right – and the NSW Right – but frontbench MPs Chris Bowen and Tony Burke are the most influential in New South Wales and co-leaders of the group. In Queensland, Jim Chalmers and his ally Anthony Chisholm are key players, too. But probably the most influential national figure is South Australian Don Farrell, a veteran senator dubbed the Godfather, who has been the leader of MPs aligned with the Shoppies (the SDA or Shop, Distributive and Allied Employees' Association) grouping within the caucus for more than a decade.

Although his power base is the six Right-aligned MPs in South Australia, he convenes the Right-aligned small state MPs (WA, SA, Tasmania, and the NT) and can count on them for influence.

And then there is the Victorian Labor Right (home to Hawke, and former senior ministers and powerbrokers Gareth Evans and Robert Ray), which, much like the Liberals, is a personalities-driven maze of sub-factions. The simplest way to explain the Victorian Right is as two major sub-factions, the Bill Shorten group or 'Shorts' and the Stephen Conroy group, or 'Cons', with a Shoppies overlay. For years, the Victorian Right has been run collectively by the ShortCons. The sub-factions work well together most of the time in Canberra, where they present a united front, but are often at loggerheads at a state level. Conroy retains influence but Richard Marles has joined the former senator as head of one of the two sub-factions in Canberra – yet Shorten has a few more supporters than Marles.

The Left is more national and therefore more straightforward. While conveners Julian Hill, Tim Ayres and Sharon Claydon have some influence, Anthony Albanese as the leader (and a long-time Left powerbroker) sits at the top of the factional pile. Tanya Plibersek is one of the most senior figures in the Left, popular with rank-and-file members and touted in some quarters as a better option than Albanese for leader. But she's less popular in the Left caucus where the leadership is decided.

South Australians Penny Wong and Mark Butler have Albanese's ear and considerable influence, as does the Victorian Andrew Giles and NSW's Pat Conroy. Queensland senator Murray Watt is close to the leader, too, as is Tasmanian senator Carol Brown, nicknamed the Godmother for her control over that state.

A splinter group known as the Industrial Left, mentioned earlier, a rump of four MPs that was led by Senator Kim Carr (who has been at loggerheads with Albanese for years and who backed Shorten for leader in 2013 and 2016) has effectively ceased to exist as a parliamentary grouping – though it still has power and influence through its alliance with major unions, in the state Labor party and labour movement. Victorian Brendan O'Connor has links to this group, too.

Michaelia Cash has filled some of the vacuum left by Cormann's exit in Western Australia while Payne and Anne Ruston – as senior cabinet ministers – have influence and moral authority, too – but among women in their party, they are the exception rather than the rule. For female Labor MPs it's a similar picture to those in the Liberal Party: Wong, Plibersek, Katy Gallagher, Kristina Keneally, Linda Burney and Catherine King are senior women who have influence and authority by dint of their roles as senior members in the shadow cabinet and their standing and popularity within the broader ALP.

Wong, in particular, has sway in terms of who, from her Left faction, is put forward for the frontbench, and also takes a strong interest in who is appointed as a deputy chair of various parliamentary committees. These women don't exactly fit the old-school mould of Labor powerbrokers sitting around a lazy Susan at a restaurant on Sussex Street, sharing san choy bao, safe seats and Senate spots (although Brown, who is not well known outside Tasmania, is more of a behind-the-scenes powerbroker in the old-school mould). Indeed, in 2021, the idea of Labor powerbrokers handing out safe seats in smoke-filled backrooms is, itself, a bit of a cliché from a bygone age.

What's the Monkey Pod lunch group?

For the Liberals, the Prayer Group, the Monkey Pod lunch conservatives, Frydenberg's Victorian group, the Rural and Regional Liberals, the Veterans group and the Morrison-Hawke Centre Right in New South Wales, as well as the aforementioned New Guard Moderates, are all sub-groups or fractions to greater or lesser degrees.

The Prayer Group has, at its core, the Prime Minister and some of his key allies including Irons and Robert. While some of its members are Pentecostal Christians, including the PM, it's cross-denominational – there are also Catholic members, plus Julian Leeser, who is Jewish. The group also includes members of the National Right such as Andrew Hastie, Amanda Stoker and Jonathon Duniam – underscoring the confluence of interests among the National Right and the Centre Right. In all, about half of the Prayer Group belongs to the Centre Right while the other half is in the National Right. And yes, they do actually get together during parliamentary sitting weeks and pray, a style of worship that some of the Catholic members of the National Right find a little strange.

The Monkey Pod, named after the tropical hardwood tree table in a meeting room in the ministerial wing of Parliament House, is a group of like-minded National Right conservatives convened by Dutton, who talk policy and tactics over takeaway lunch on Tuesdays. The group had been meeting for years but it was in November 2015 – a few months after Malcolm Turnbull became prime minister and when Tony Abbott started attending the meetings – that they shot to prominence and were labelled a 'resistance movement' backing the recently deposed PM. Abbott, famously, took a chocolate cake baked by his prime ministerial chief of staff Peta Credlin – herself a lightning rod for discontent during his prime ministership – to one of the lunches.

The Rural and Regional grouping of Liberals, convened by South Australian Rowan Ramsey and with at least 19 members, advance regional Liberals' interests in contra-distinction to the 21 Nationals MPs, rather than being a formal faction. And Frydenberg's 'ambition faction' has ten members, some formerly Moderates and some from the National Right. They meet semi-regularly when in Canberra and work to advance the state's interest and ensure their preferred Victorian candidates are preselected.

For Labor, mostly, sub-groupings and other alliances are generally tied to the union movement. Aside from the Shoppies, other major unions aligned with the Right faction include the Transport Workers' Union and the Australian Workers' Union. On the Left, it's the Australian Manufacturing Workers' Union, the United Workers Union, the Community and Public Sector Union and the powerful CFMEU. Non-union groups such as the Labor Environmental Action Network and Open Labor are activist groups that have some influence on policy, too.

What if you don't want to be in a faction?

Some MPs simply don't want to be in a faction, although that's uncommon, particularly in Labor. Three non-aligned MPs from the ALP – Bob McMullan, Andrew Leigh and Alicia Payne – have all represented the ACT, which has a small, well-educated population with active branches that don't like having candidates imposed on them by head office.

Then there are the independents in the Liberal Party. These include House Speaker Tony Smith and Senate president Scott Ryan (both of whom belonged to the Victorian Costello faction, when it existed) as well as WA senator Dean Smith – a conservative but also a leading proponent of marriage equality who now tells colleagues he is a 'faction of one'. Fellow West Australian Liberal Celia Hammond straddles all three groups – she was backed by the National Right's Cormann to take Julie Bishop's seat, is a member of the Prayer Group and holds progressive views on climate change.

These independents are a reminder that in the Liberal Party, in particular, some MPs simply don't want to be pigeonholed, and that over time, alliances and allegiances can change as philosophies and interests evolve.

FEDERAL FACTIONS (AND FRACTIONS) AT A GLANCE

| Labor powerbrokers | | Liberal powerbrokers | | | Liberal subfactions (key members) | | | | | | |
LEFT	RIGHT	MODERATES	CENTRE RIGHT	NATIONAL RIGHT	NEW GUARD MODERATES	MORRISON CLUB	PRAYER GROUP	FRYDENBERG 'AMBITION' FACTION	MONKEY POD LUNCH CONSERVATIVES	RURAL & REGIONAL	VETERANS
Anthony Albanese (NSW)	Chris Bowen (NSW)	Simon Birmingham (SA)	Scott Morrison (NSW)	Peter Dutton (Qld)	James Stevens	Scott Morrison	Scott Morrison	Josh Frydenberg	Peter Dutton	Rowan Ramsay	Stuart Robert
Penny Wong (SA)	Tony Burke (NSW)	Marise Payne (NSW)	Josh Frydenberg (Vic.)	Michaelia Cash (WA)	Andrew Bragg	Alex Hawke	Andrew Hastie	Greg Hunt	Zed Seselja	Dan Tehan	Andrew Hastie
Mark Butler (SA)	Bill Shorten (Vic.)	Paul Fletcher (NSW)	Ben Morton (WA)	Zed Seselja (ACT)	Jason Falinski	Ben Morton	Stuart Robert	Michael Sukkar	Michael Sukkar	Warren Entsch	Jim Molan
Andrew Giles (Vic.)	Don Farrell (SA)	Trent Zimmerman (NSW)	Alex Hawke (NSW)	Michael Sukkar (Vic.)	Tim Wilson	Stuart Robert	Amanda Stoker	Alan Tudge	Angus Taylor	James McGrath	David Fawcett
Pat Conroy (NSW)	Richard Marles (Vic.)	Jason Falinski (NSW)	Stuart Robert (Qld)	Angus Taylor (NSW)	Jane Hume		Alex Hawke		Alan Tudge	Angus Taylor	
Kim Carr (Vic.)	Jim Chalmers (Qld)	Andrew Bragg (NSW)	James McGrath (Qld)	Andrew Hastie (WA)	Trevor Evans				Andrew Hastie		
Tanya Plibersek (NSW)	Joel Fitzgibbon (NSW)	James Stevens (SA)	Greg Hunt (Vic.)	Tony Pasin (SA)							

Factions map of federal Liberal and Labor parties, 46th Parliament, 2021

8

WHAT IS LOVE AT FIRST SIGHT?

Can a chemical reaction in your brain
really decide your romantic future?
Or have we been sold a fairytale?

Julia Naughton

It happened in an instant: our eyes met across the dance floor and we moved towards one another. Something pulled us closer together.

'What's your name?' I asked with the confidence of a 24-year-old who had zero expectations of meeting anyone, let alone a romantic partner, at a rave party at 3am on New Year's Day.

'Nick,' he replied. 'What's yours?'

It was his eyes and then his smile that drew my body closer to his. We danced until the sun came up. Then we went our separate ways and, as with all great millennial romances, a friend request on social media followed. By April, we were Instagram official.

Had the universe been turning its gears, preparing us both for that moment on the dance floor? Did the stars finally align? Perhaps it was simply potluck. All I know is that seven years later, we have married and, yes, we are very much in love.

My experience is far from unique. Poetry, novels, music – all kinds of art – have popularised the phenomenon of love at first sight for centuries. But what is it, how do we measure it? Does it even exist? What happened on the dance floor that night? And was a romantic relationship inevitable?

*Poetry, novels, music – all kinds of art –
have popularised the phenomenon of
love at first sight for centuries.*

What is love at first sight?

One hundred milliseconds – that's precisely how long it takes to evaluate a potential sexual partner. And if, after that split second, you deem that person to be attractive, emotions will begin to kick in as well as some animal-like instincts.

Anthropologist Helen Fisher explored the process of communicating our attraction to another person via a copulatory gaze in her book *Anatomy of Love*, first published in 1992. 'The gaze is probably the most striking human courting ploy,' Fisher writes. In cultures where eye contact between the sexes is permitted, 'men and women often stare intently at potential mates for about two to three seconds during which their pupils may dilate – a sign of extreme interest. Then the starer drops his or her eyelids and looks away.'

So what is it that makes our pupils dilate? What is it that triggers a snap judgement that could lead to a romantic connection?

Unsurprisingly, looks matter. Studies reveal we are sexually attracted to people who look like us. In 1999, researchers asked subjects to rate pictures of faces morphed with their own and found the subjects each rated the morphed faces as being more attractive. Other physical traits can also seal the deal: 'Men and women around the world are attracted to those with good complexions. Everywhere people are drawn to partners whom they regard as clean. And men in most places generally prefer plump, wide-hipped women to slim ones,' Fisher writes.

The medial prefrontal cortex, near the front of the brain, is responsible for such judgements. Researchers in Dublin discovered that one particular area of this region – the rostromedial prefrontal cortex, a segment located lower in the brain – goes one step further in evaluating physical attractiveness by asking, 'Is this person a good match for me?' This is the same brain region known to be important in social decisions, particularly how similar someone is to you, the study's authors explained. Essentially, we are evaluating someone's attractiveness in those initial milliseconds while also determining their compatibility.

But it's all very unromantic, really, says Trish Purnell-Webb, a clinical psychologist at the Relationship Institute Australasia.

'When we're looking to mate, the brain will constantly search the environment and look for another person who will be a good mate and make strong babies.'

You like what you see, so what happens next?

There is a surge in the attachment hormone oxytocin, says Purnell-Webb. Oxytocin stimulates the secretion of other neurotransmitters – molecules that act as chemical messengers – such as dopamine which fires up the brain's reward centres.

At this point, you begin to focus more intently on that person and there will be a spark of interest in your eyes. 'Imagine a dog pricking up its ears when it senses its master has returned home – that's what a person who is experiencing attraction is like,' says Purnell-Webb, adding that generally it's subtler than with dogs, although not always.

Often called the 'cuddle hormone', oxytocin compels us to get close; when we are feeling close to somebody else, we secrete more of it. 'Oxytocin is actually the hormone for poor decision-making,' says Purnell-Webb. 'It makes you become a bit obsessed and compulsive. You're being driven to try and make contact and attract the same interest back.'

The more signs you get from the other person that they are interested in you, the more oxytocin is released. It's during this time that we often feel like our heart is racing, our palms get sweaty or we have butterflies in our tummy, says Gery Karantzas, the director of Deakin University's Science of Adult Relationships Laboratory. 'It's why we often hear love being referred to as a drug,' says Karantzas. It's a virtuous circle and a delicious cycle. 'You develop these strong feelings for someone, particular parts of your brain get activated when you spend time with them, you therefore want to spend more time with them, which then releases more of these chemicals, and you feel more of those feelings and so on.'

Similar brain areas fire up if you take cocaine, as Fisher discovered in a groundbreaking experiment: she did MRI scans on 37 people who were madly in love, which showed a surge of activity in brain areas rich in dopamine. 'But romantic love is much more than a cocaine high – at least you come down from cocaine,' Fisher said in a 2014 talk. 'Romantic love is an obsession, it possesses you. You lose your sense of self.' You can't, she adds, stop thinking about that other human being.

What's love at first sniff?

Pheromones – which are, essentially, scented chemicals that act like hormones outside the body – allow us to literally sniff out a potential match. Think of them as a primal form of communication; smell plays a major role in sexual attraction and can even trigger ovulation. But here's the thing: pheromones smell different to different people. 'Male pheromones are often described as having a woody or musky smell while female pheromones smell more floral or sweet,' says Purnell-Webb. At other times, depending on who is sniffing whose pheromones, they can smell bitter, sweaty or like stale urine. Some studies have suggested that, during ovulation, women are highly sensitive to the musk-like pheromones that men secrete.

As you continue to smell the other person, attraction builds and you move closer to them. Are you looking for anything in particular? Not really. It's more a case of beauty being in the eye of the beholder, says Purnell-Webb. 'We all have particular personal preferences, some people like beans, others peas,' she says. There is some evidence to suggest attraction is part of our genetic imperative – our genes determine who and what we are attracted to – which raises the question, is there a perfect match out there for each of us?

That's the premise of the Netflix show *The One*, set 'five minutes into the future', in which finding a soulmate is as simple as snipping off a strand of hair and posting it to a DNA matchmaking company. Because the company claims to find perfect matches on a genetic level, attraction is inevitable, actually involuntary. Upon first meeting, the match smells right, looks right and demonstrates all of the right characteristics. They also taste really good.

A 2000 study by Swiss biological researcher Claus Wedekind found that when we kiss we exchange a heap of biological and genetic information. Women tend to be more attracted to men whose MHC (major histocompatibility complex) genes, which are essential to the immune system, are different from their own. Having different genes produces offspring with stronger immune systems. Wedekind is best known for his 1995 'sweaty T-shirt study', which had a similar upshot but involved women sniffing T-shirts that had been worn for two days by male students using no deodorant – they preferred the odour of men with different MHC genes to their own.

No one has conducted a T-shirt-sniffing exercise to examine how same-sex participants are magnetised towards some people but not others. However, a large genetic study into the underpinnings of same-sex attraction, whose findings were released in 2018, had researchers identify four regions in participants' genomes that influence a person's choice in sexual partner. Two were observed in men and women, and two were seen in men alone. The DNA identified could account for only 8 to 12 per cent of the genetics behind non-heterosexual behaviour. One of the variants was linked to the olfactory receptor.

Sniffing unwashed T-shirts is the human equivalent of dogs sniffing each other's sweat glands (which happen to be located in their rear ends), says Purnell-Webb. Romantic, no? And kissing, for its part, may well be the most glamorised exchange of chemical information known to humans. It also explains why some people are immediately turned off after the first whiff (or kiss) while, for others, the intensity of attraction and desire only grows.

Can you engineer love at first sight?

Researchers of romantic love will tell you there are some definitions of love that would allow for the possibility of love at first sight, while other definitions leave no such room.

If we think of love as a purely physiological experience, then love at first sight is possible, says Mandy Len Catron, who wrote a book called *How to Fall in Love with Anyone*. 'If we looked at what's happening in your brain when you are feeling an intense attraction to a complete stranger, we'd see dopamine and norepinephrine levels rise and we could definitely say you are having a love experience,' she says.

Of course, this definition has several implications. For one, it potentially removes personal responsibility from the act of falling in love. You're simply going along for the ride and then – bam! – destiny steps in and offers up your one true love.

But if we were to think about love from a sociological or philosophical perspective, says Catron, it becomes less about a *feeling* in our bodies that happens to us, and more about an *action* we choose. In this case, love is a verb – and even if we're attracted to someone, we can still decide they're not for us, for a whole range of reasons. By this reading, love requires work and doesn't always fit so rosily in the love-at-first-sight wonderland, where individual effort is void and soulmates are served on silver platters.

Still, we gleefully consume and re-tell stories of being instantly struck with love, each tale another hit of ecstasy that fortifies the possibility that one day we might find extraordinary love. 'I understand why love at first sight is appealing,' says Catron, whose 2015 essay for the *New York Times'* Modern Love column, titled 'To Fall in Love with Anyone, Do This' catapulted a 20-year-old experiment into the cultural lexicon. The 1997 study, by US psychologist Arthur Aron, succeeded in making two strangers fall in love in a laboratory by having them answer a series of

36 personal questions face to face – What would constitute a 'perfect' day for you? For what in life do you feel most grateful? – and then having them stare into each other's eyes for four minutes. Six months later, the two strangers were married.

Catron, who came across Aron's work while at university, decided to apply his technique in her own life, with an acquaintance. Naturally, they weren't in a laboratory but a bar and they weren't strangers. Nevertheless, it worked. Three months later they were still together and Catron did what any writer would do: she wrote an essay about it. Published just as dating apps such as Tinder were sweeping the northern hemisphere, the piece offered something powerful, says Catron: a 'ready formula for falling in love' that contradicted the idea of love at first sight, that love was something that simply happened to you. After all, Catron chose to do the experiment; the acquaintance agreed. It was through these actions – and perhaps Aron's carefully selected questions – that they were able to achieve and sustain intimacy.

Speaking on the phone from her home in Vancouver in 2021, Catron says she is still in a relationship with that university acquaintance from her experiment six years ago. Meanwhile, our fondness for the idea of love at first sight remains largely unchanged across cultures. In arranged marriages, for example, couples report similar feelings, only it is present during the marriage as opposed to the courting period. 'The reality is that, by nature, humans are very interested in story,' says Catron. 'We have evolved to be narrative creatures and when it comes to romantic love we really like the idea that there is a soulmate or that there is someone out there who we are favoured to be with.'

There are countless examples – both fictional and real life – that have been associated with the notion of love at first sight. In Disney's *Sleeping Beauty*, Princess Aurora meets Prince Phillip for the first time on her sixteenth birthday and they instantly fall in love. In William Shakespeare's *Romeo and Juliet*, the star-crossed lovers first lay eyes on each other at a ball – neither of them speak – yet after learning of Juliet's name from a waiter, Romeo becomes infatuated, immediately declaring he has never been in love until this moment! It's all very quick and erratic. And it's not just Shakespeare; Kylie Minogue's 2001 hit 'Love at First Sight' tells the same story, as do countless other pop tunes.

Can you make love at first sight feelings last longer?

According to Purnell-Webb, between nine and 12 months is the limit before the butterflies calm down. That's the amount of the time it takes for a female to become pregnant and for the foetus to have developed to a point where she is likely to have a live birth. 'It's all about keeping the human race going,' she reminds us.

But Karantzas says passionate love ebbs and flows in a relationship over time. 'We see an increase in other types of love as relationships go along,' he says. Companionate love, for example, where having a deep sense of caring and concern for someone, often grows with a relationship.

And when you register someone as your 'lifelong love' the moment you see them, only time can tell whether you're right or not. 'People will say, "Well, I knew instantly that I loved this person" but often the people who are saying that are those who have been together five, 10, 15, 20 years.'

Purnell-Webb says, 'Attraction is what happens, Hollywood romanticists have labelled attraction "love at first sight" but it doesn't really exist, love only develops over time.'

Catron prefers to align her ideas about love with writer Bell Hooks' definition – that is, love is an action, never simply a feeling.

'It's not about how you feel, it's about what you do with those feelings and how you connect to another person and how we treat each other,' she says. 'By defining love as an action, we are more likely to have healthier, more loving relationships that are sustainable.'

9

WHO RUNS CHINA?

To join the century-old Communist Party of China is a gruelling process, yet 92 million citizens have done it. What are the benefits? And who calls the shots?

Eryk Bagshaw

From a clandestine meeting of 53 people in a Shanghai house to 92 million members across China, the Communist Party of China has driven the longest and largest political movement the world has ever seen.

In doing so, it has survived internal revolts, foreign invasions and economic devastation to wield more than 70 years in power. It has crushed internal dissent, wiped out cultural identities and overseen the fastest economic expansion of the modern age.

'Follow the Party forever,' the banners in Beijing declared on 1 July 2021, the Party's centenary.

China is now in the midst of the world's largest social experiment. It is attempting to harness all the benefits of a market economy while eliminating any capitalist threat to its rule. Can an authoritarian government remain in control while it liberalises its economy – but not its population? What role does the Party play? And will it stay in power?

> *China is now in the midst of the world's largest social experiment.*

How do you become a member of the Party?

Hassled by local police, the 53 original Party members who met in a Shanghai shikumen terrace on 23 July 1921 had to travel 100 kilometres to Jiaxing and conduct their affairs on a boat before they could elect their secretaries, organisation and propaganda chiefs. Among them was Mao Zedong, the revolutionary and dictator who would go on to found the People's Republic of China under Communist rule. It took almost thirty years of bloodshed, political manoeuvring and popular uprising for Mao and the Party to take power, but it has held it ever since.

Now, the party of the farmers is becoming the party of the middle class. Better educated, aspirational and more financially secure, it has changed as China's own wealth has grown, but it is also increasingly trying to raise the standard of its membership base.

Open only to Chinese nationals aged 18 and over, the application process is, by Western standards, gruelling. In Australia or the United States, party members can fill out little more than a three-page form to join the Liberal, Labor, Republican or Democratic parties. In China, party aspirants have to submit in writing why they should be considered. Their reasons have, historically, swayed with the direction of the leadership. Under Mao they were heavily ideological; people joined the Party because they believed in its cause. But over time, they have become more pragmatic, with applicants echoing how they can serve the Party and the country, academically and professionally.

If they get through this stage, the Party will conduct background checks on their parents, family and social connections. Once vetted, they move onto an initial three-day course on ideology, which covers the theories of three major leaders – Mao (who headed the People's Republic of China from its inception in 1949 until his death in 1976), Deng Xiaoping (1978 to 1989) and Xi Jinping (since 2012) on 'socialism with Chinese characteristics' – before submitting papers on current events every three months. Interviews will be conducted with eight friends and family to verify the applicant is ideologically pure.

At this stage, they are allowed to officially apply – but the process is not over. Quarterly interviews track their progress during a year-long trial membership before they are admitted finally. 'The vetting process has become much more difficult,' says Mercator Institute for China Studies analyst Nis Grunberg. 'Once you're in there, you need to continuously prove yourself.'

More than 90 million Party members have gone through this process. That is more than the number of Republicans and

Democrats combined. It is a lot of people – but still less than 7 per cent of China's 1.4 billion population. Of the world's political parties, only India's Bharatiya Janata Party has more members. Up to 80 per cent of CCP applicants are rejected and have to reapply. Increasingly, the base is skewing towards members with academic degrees. Figures from the Party's Central Organisation Department analysed by Grunberg for Mercator show the ratio of workers and farmers has fallen from almost 40 per cent in 2009 to a third in 2019. At the same time, the number of members with a university education has risen from just over a third to half.

Businesspeople were not even allowed to join the Party until 2001. But the shift to more university-educated members has been tied to a drive into the private sector. Of the 15.61 million private companies in China, 73 per cent now have a party cell installed. The cells are a key part of the CCP's attempts to steer market-driven innovation while ensuring no entity becomes more powerful than the Party.

Internally, the cells are filled with a mixture of ideologues, pragmatists and ambitious members using the Party to bulk up their resumé. 'It is difficult to make sure that everyone who's enjoying the benefits of membership really believes in his or her heart that Marxism is the best way forward,' says Grunberg. 'That's a long stretch for a person who grew up in Shanghai in the 2000s. But at the end of the day, it doesn't really matter if you strongly believe in the Party. If you act upon the rules and recreate them on a daily basis, they are going to stay.'

What are the obligations and benefits of membership?

The Party very much focuses on unity and cohesion, says Adam Ni from the China Policy Centre, an independent, non-profit research organisation based in Canberra.

'You're demanding a lot from your members. It's not a soccer club. It's something that you dedicate your life to. It's also a vehicle

It's something you dedicate your life to. It's also a vehicle for upward social mobility in China.

for upward social mobility in China. So, if you're somebody ambitious, if you're smart, then it's the best way to move up the social ladder in China, because it's the organisation that monopolises political power.'

Out of that political power come economic opportunities and access to political, business and academic networks. Jobs in the prestigious civil service or the $30 billion state-owned enterprises sector are overwhelmingly skewed towards Party members.

According to the Party's constitution, members have a duty to attend meetings, keep up to date with political documents and participate in discussions through the Party's journals and newspapers. Historically, members have had limited rights to criticise the Party's organisation but this has become more dangerous under President Xi Jinping, with control increasingly centralised around the president and closely monitored by internal surveillance networks.

The oath taken by members when they join the Party reads:

It is my will to join the Communist Party of China, uphold the Party's program, observe the provisions of the Party constitution, fulfil a Party member's duties, carry out the Party's decisions, strictly observe Party discipline, guard Party secrets, be loyal to the Party, work hard, fight for communism throughout my life, be ready at all times to sacrifice my all for the Party and the people, and never betray the Party.

In the private sector, Party cells can vary from zealous internal workplace surveillance units to hour-long, tick-box discussions on socialism with Chinese characteristics before everyone knocks off for drinks. Ryan Manuel, a former director of policy research at the University of Hong Kong and Australian government adviser who now runs the Official China research firm, says this achieves two things.

'The first is a traditional Chinese focus on inculcating moral values under the guidance of the Party, rather than seeking checks and balances on individual power. The second is ensuring that the Party has a voice in all private enterprises and continues to encourage large state-owned enterprises.'

Historian Xiao Gongqin recalled in an essay published in 2020 how China had learnt its lessons from the fall of the Soviet Union, which had undermined its state power by decentralising its economy.

'This man looks smart but he is actually stupid,' former Chinese president Deng remarked after meeting the last leader of the Soviet Union, Mikhail Gorbachev, in 1989.

'Gorbachev looked good,' Xiao later wrote. 'But he forgot that the leadership of the Communist Party is the basis for stabilising the entire social and economic development. Abandoning the leadership of the Communist Party is actually detrimental to the political stability required for economic development.'

How does the Party work?

Above the 92 million members are five key bodies. The Party Congress selects its 2354 delegates from the membership base, and they attend the annual political gala of the National People's Congress in Beijing to rubber-stamp key laws and leadership positions such as the Party general secretary.

Next is the Central Committee, made up of 205 members and 171 alternate members, who can participate in policy plenums but can't vote. The alternate members are a feeder club for the Central Committee: members can be voted into full committee status if full members retire or are expelled. The committee itself meets formally once a year to discuss policy and it oversees various executive national political bodies. The committee also, in theory, elects the next level of authority, the Politburo but, in reality, these positions are decided by powerbrokers behind closed doors.

Party General Secretary
XI JINPING

Politburo Standing Committee
7 MEMBERS

Politburo
25 MEMBERS

Central Committee
205 MEMBERS & 171 ALTERNATES

Party Congress
2354 DELEGATES

Communist Party of China
92 MILLION MEMBERS

1.4 BILLION PEOPLE
China's population

Infographic by Matthew Absalom-Wong

The 25 members of the Politburo are a mixture of representatives from the Central Military Commission, ministers for development and reform, regional leaders and discipline chiefs. Vice-Premier Sun Chunlan, who led the internal intelligence and propaganda Central United Front Work Department until 2017, is the only female member of the Politburo.

Within the Politburo, the seven-member Standing Committee is the key decision-making body. It includes Premier Li Keqiang, anti-corruption chief Zhao Leji and national security director Li Zhanshu. Above all of them is Xi.

'One of the features of the party structure is the concentration of power as you go up the structure,' says Ni. 'It's a Leninist organisation in the sense of its organisational discipline.'

That centralisation is becoming much more top heavy under Xi, reaching right down to the village level. There are up to a million villages across China. 'Historically, townships and villages have had nearly total autonomy, in practice, over their affairs,' says Manuel. But since 2018, the Party has forced villages to report on how finances are being spent each year. This allows it to make sure townships are using revenue to meet targets from above. 'And if you don't make them, then we're going to send down the anti-corruption people,' says Manuel. 'It is a credible threat.'

The same is true of provinces scrambling to work out how they are going to help meet Xi's target of net-zero emissions by 2060. Many of them still have coal-fired power plants being built and some will be threatened with not meeting new Party standards if they do not mothball them. 'It's the watermelon problem,' says Manuel. 'If we want a green China, we have to accept a red China to enforce a goal.'

Lowy Institute senior fellow Richard McGregor says China is a large, diverse and unwieldy country. He recalls a Chinese colleague once saying to him: 'In China, people will have to fear the government, otherwise they won't respond to it.'

Yet it has achieved what many others could not: half a century of continuous economic growth. China has, in effect, doubled its GDP every eight years since liberalisation in 1979–80, lifting 800 million people out of poverty, which is like nothing history has ever seen before. From being a backward nation with poor infrastructure, China has become the world's second-biggest economy, biggest manufacturer, biggest merchandise trader, and largest holder of foreign exchange reserves. 'China has some very smart policymakers who have a very good track record,' says McGregor.

Manuel says Xi has taken powers from local leaders and executive bodies and given them to the legislatures and internal inspectors in a sort of top-down populism that forces people to follow his orders more strictly. 'They care a lot about what the next person thinks,' he says. 'Everyone is always competing for the favour of the people above them. Given there are over 90 million party members across more than 30 provinces, nearly 900 municipalities and nearly 3000 counties, there exists a vast bureaucracy that is fundamental to the prosecution of the leader's interests.'

> *Everyone is always competing for*
> *the favour of the people above them.*

The outcomes can be good for policies that are likely to deliver positive outcomes, such as emissions targets, but brutal for others, such as the repression of the Uighurs and other ethnic minorities in Xinjiang or the crackdown on the pro-democracy movement in Hong Kong.

'It's influencing every walk of life in Hong Kong,' says former Hong Kong legislator Ted Hui, who left the former liberal enclave while on bail facing national security charges and now lives in Australia. 'The CCP is pushing it to an extreme, basically declaring that any dissent and opposition operating in Hong Kong will be harmful to national security. The new National Security Bureau has no checks and balances by the judiciary and there's no transparency. So, basically, they can do whatever they want.'

Is the Party's power limited?

China's deputy ambassador to Australia, Wang Xining, took issue with the West's references to the Chinese Communist Party in April 2021. 'It's the CPC – the Communist Party of China,' he told the National Press Club. 'Not the CCP, the Chinese Communist Party. There are different linguistic connotations. There's a very shallow understanding of the role of the Communist Party of China.'

In Wang's formulation, the Communist Party of China belongs to China and its people first. But, as the West would have it, the past few decades have shown the Party is doing what it can to ensure China's workers, businesses and leaders belong to it.

McGregor, who wrote *The Party: The Secret World of China's Communist Rulers* a decade ago, says that while historically the Party set policies, and the government, civil service and courts executed them, this distinction has been effectively abolished under Xi. 'Which just means it's a much more controlled environment with less room to move on policy issues and policy debates,' he says. 'Look at the Politburo Standing Committee and their tasks. A normal government has a minister of finance, environment, foreign policy. If you look at the Standing Committee, virtually none of them have an executive function. They are all political jobs involving ideology and anti-corruption.'

Manuel observes, 'It is like having more stuff done by the Liberal Party backroom rather than done by the public service, by ministerial staff or by legislators.'

Grunberg says the Party and state are being fused tighter than before. 'What used to be Party rules have been made into laws,' he says.

In the 1980s, Deng Xiaoping's Central Committee preferred not to mention the Party in the main text of the People's Republic of China constitution, arguing that as Party members represented only 5 per cent of the population its role was best left unsaid. But under Xi, Party leadership has now been written into Article 1 of the constitution's main body as the 'defining feature of socialism with Chinese characteristics'. At the same time, the pretence of state laws being separated from Party directions has been whittled away as the Party's role has been written into state governance, turning internal Party ideological, organisational and operational breaches into legal ones.

Since 2017, four key developments have demonstrated how this party-state fusion is creating a minefield for governments and companies. In June of that year, the Chinese government implemented the Cybersecurity Law, which requires data stored within China, including by foreign companies, to be handed over for government security checks. In the same month, a new National Intelligence Law enabled the compelling, without a warrant, of any 'organisation or citizen to support, assist and cooperate with the state intelligence work'. The law was identified as a key reason behind Australia's decision to block telecommunications company Huawei from Australian's 5G network in 2008.

Then, in September 2020, the Central Committee outlined its plans to increase its ideological power over private companies through its United Front network, by closely uniting business figures 'around the Party' and strengthening the ideological monitoring of workers through internal company Party committees.

One of China's richest men, Jack Ma, the founder of the Alibaba online retail empire, soon fell foul of the Party's tightening grip. After a rapid expansion into online banking platform Alipay, now used by 70 per cent of China's population, Ma criticised the Party and China's regulators for stifling innovation. 'We shouldn't use the way to manage a train station to regulate an airport,' he said. 'We cannot regulate the future with yesterday's means.' Ma's Ant group had been preparing for the world's largest share offering but the estimated AU$400 billion market launch was scuttled by Chinese regulators. *The Wall Street Journal* reported it was on the orders of Xi himself. Ma promptly disappeared from public view for three months, his whereabouts unexplained, before resurfacing in January 2021. 'My colleagues and I have been studying and thinking, and we have become more determined to devote ourselves to education and public welfare,' Ma said.

The next month, Xi gave a speech on the country's achievement in eliminating absolute poverty. The milestone is a key marker on China's path to rejuvenation, as outlined by the Party over the past century. Throughout his speech, Xi bound together the Party, the nation and its people. He showcased the mobilising power of nationalism and the socialism with Chinese characteristics that is increasingly becoming Xi and China's ideological trademark. 'The arduous task of eradicating absolute poverty has been completed, and another miracle has been created in the annals of history!' he said. 'This is the great glory of the Chinese people, the great glory of the Communist Party of China and the great glory of the Chinese nation!'

Willy Lam, a political analyst with the Chinese University of Hong Kong, says in China, 'nationalism is one of the key pillars of legitimacy and the leadership wants to show the people that they can stand up to the pressures from the Western world.'

Xiao, the veteran scholar of the Party, said in an essay published in December that 'China has gotten rid of the ultra-leftism and leftism'. 'The ideology of China, on the basis of respecting common sense, respecting the diversity of social economy and culture, and respecting the historical continuity of the existing order, transcends the left and right,' he wrote. 'Xi's goal is to make the

Xi's goal is to make the whole world see China as a great power, and him as a key figure in making it great.

whole world see China as a great power, and him as a key figure in making it great. At heart, he's a nationalist.'

Is Xi really the Chairman of Everything?

Xi, the son of a guerrilla revolutionary, is now president for life. He was exiled with his father during Mao's purge in the Cultural Revolution. The story goes he lived in a cave and was rejected for Party membership nine times before being accepted as a 'Worker-Peasant-Soldier student'. He would go on to finish a degree in chemical engineering and a doctorate in law and ideology from Tsinghua University.

> *The story goes he lived in a cave and*
> *was rejected for Party membership*
> *nine times before being accepted.*

Xi moved steadily up the ranks for two decades. Critically, he replaced Chen Liangyu, the party secretary of Shanghai, in 2007 after a corruption scandal, putting him on a path to the Politburo Standing Committee.

By 2012 he was president. Over the next six years he would consolidate his power through more corruption purges, and increase China's domestic and foreign ambition. Two key policies are driving this further: dual circulation, which will see state-backed domestic production among China's 1.4 billion people become the dominant source of economic growth in the export-driven powerhouse; and the Belt and Road Initiative, a AU$1.5 trillion dollar infrastructure push across more than 100 countries in Asia, Africa and the Indo-Pacific.

When it was time to nominate Xi's successor in 2018, there were no clear challengers to his authority. He removed the limit of two five-year terms instituted after Mao, which had been put in place to avoid a cult of personality forming around a leader.

Carl Minzner, a professor of Chinese law and politics at Fordham Law School, says there is a narrative that is marginalising China's other former leaders Hu Jintao, Jiang Zemin and Deng together as it raises the profile of Xi. 'It starts down the path of presenting post-2012 [after Xi became president] as the culmination or solution to millennia of Chinese history,' he says. 'I think that the name of the game right now is raising the ideological profile of Xi Jinping himself up to something quite close to, or surpassing that of, Mao himself.'

It is a message being reinforced in Chinese state media daily. On one day in April 2021, all six of the front-page stories in the *People's Daily* newspaper were about Xi. In new textbooks distributed throughout China for the centenary of the Party, Minzner notes that *On the History of the Chinese Communist Party* contained 40 speeches by Xi, or 180,000 characters, while excerpts from Mao, Deng, Jiang and Hu were worth a total 98,000 characters.

As Ni says, this streamlined narrative is reinforced through Party messaging: 'There is a straight line between Mao and Xi. Where Mao led the standing up of China, Deng Xiaoping led the development of China and now Xi is leading China to become a nation of strength.'

Strongman rule wasn't always so popular. Former Central Party School professor Cai Xia told Radio Free Asia (RFA) in 2020 that for a while, between 2001 and 2006, there were active discussions about 'intra-party democracy' to solve issues of political system reform in China and to promote the market economy. When it was suddenly announced in 2018 that the two-term limit would be abolished for Xi, members of the Party who had thrived under Hu were shocked. Notably, Xi did not change the retirement age limits for any other high office, effectively eliminating any direct competition.

'He forced everyone at the [Plenum] to swallow the revision like he was stuffing dogshit down their throats,' said Cai in a leaked speech to a private group in June 2020. 'So many Central Committee members were at the session, yet not one dared to raise this issue.'

Cai is now in exile in the United States. Many of her colleagues who have criticised Xi have also either left the country or been detained. 'Xi Jinping is calling all the shots on major issues. I call him a gang boss because there is no transparency, and there is no decision-making mechanism. When different opinions surface, coming from people like me, they can expel you from the Party and take your pension away,' Cai told RFA in August 2020. 'He has a tight grip on everyone. The advanced surveillance technology is not only utilised in monitoring Xinjiang and Tibet but it is also applied to monitor CCP members as well as mid- and high-level officials. Around 2013, Xi Jinping also announced a policy that forbids the formation of any alumni associations or hometown associations; additionally, gathering after work is also not allowed. He was worried that such gatherings may provide room for cliques or political factions to grow within the Party.'

When different opinions surface . . .
they can expel you from the Party
and take your pension away.

Manuel says someone always loses in any political upheaval. In the rise of Xi, it's the second-generation elite such as Cai and their families who have been either forced into silence, hiding or exile, leaving Xi unchallenged at the top of the CCP pyramid.

'These are people who have gone to Harvard or Yale, who speak excellent English, and they don't like Xi.'

He says the combination of the Party as an ideological commitment and as a vehicle for professional promotion has left this group of potential Chinese leaders sidelined.

'These people are seeing their purpose torn up,' he says. 'Xi Jinping doesn't like that group of members, he likes true believers because he's a true believer.

'And that really makes it harder for us to deal with China.'

What happens after Xi?

Four scenarios for a change of leadership have been outlined by McGregor and Jude Blanchette from the Center for Strategic and International Studies, a bipartisan Washington think tank affiliated with Georgetown University whose focus is on US interests and security. They are as follows: an orderly transition in 2022; a succession plan to retire at the twenty-first Party Congress in 2027; a leadership challenge; or unexpected death or incapacitation.

China is big on targets and anniversaries. In 2021, Xi claimed the nation had eliminated absolute poverty. Now at the forefront of Xi's mind will be China's next goal, achieving 'modern socialism' by 2035, reaching the level of 'moderately developed countries'.

By 2035, Xi will be 82, the same age as Joe Biden at the end of his first term.

'I think he'll be around for a long time,' says McGregor. 'By 2035, which I think is a midpoint for the China dream, Xi will be 82, which is the same age as Joe Biden at the end of his first term. So, I think we should expect him to be around for at least another five, six years to 2027. And probably beyond that, if not leading the party then running it from behind the scenes.'

And while China is marking a century of its Communist Party, one big anniversary remains.

That date is 2049. It falls a century from the founding of the People's Republic of China by the Communist Party after it had completed its three-decade-long transition from guerrilla fighters to rulers of the Middle Kingdom. It is also the deadline that Xi has set for China to eclipse the United States and European countries as a fully developed modern nation and to unify Taiwan and complete the 'China dream'.

In the middle of these grand historical benchmarks and power struggles are 1.4 billion people, each with their own stories, ambitions and failures.

'I think one of the ways to understand China is that it's a massive country, not just the CCP,' says Ni. 'The CCP is the ruling regime but it's a living, breathing place with a lot of people. These people are living with their hopes and fears, within limitations and prejudices and history. And with all that diversity comes complexity. I think sometimes we tend to reduce China to a geopolitical context, or seeing it through the view of the Party or competition.

'But I think it is far more complex than that.'

10

HOW DO YOU STOP CRIMES OF FIXATION?

Some offenders are fuelled by obsession – and heavy-handed policing doesn't always help. So, how do the experts stop crimes before they start?

John Silvester

The man on the skateboard is searching for trouble and he knows where to find it. He is one of those COVID protesters looking for police to film while breaching lockdown rules – and he chooses Victoria's Parliament House as his stage.

The police look a little jaded as they ask his name and address. The man's voice is brittle with anger and he seems fuelled with a self-perpetuating sense of outrage. He clearly thinks he has been wronged and is the victim.

Eventually, he tells them if they want to know his background they should contact Inspector Steve Cooper from the Fixated Threat Assessment Centre. This is a red flag to the police. They know their man is a person of interest to the people who try to stop a small number of the troubled turning from angry to violent.

As they usher him from the road and he continues to berate them, declaring they are all terrorists, it looks as if things could turn ugly. That is until a uniformed cop steps forward and addresses him by his first name. 'Do you remember me?'

In that moment, the angry man's attitude changes. He recognises the policeman as someone who treated him well when he was under arrest at a suburban station. 'Ask Chris, he'll tell you I'm a good guy,' he says, using the cop as his living reference.

That cop, at least, is no longer an oppressor but someone who can help.

Every suburb and every country town has them. The regular offenders who need friendly care and psychological treatment rather than stern words and a locked cell. Often this is managed, case by case, by individual police without a detailed game plan. Most times they battle through, sometimes they don't, with disastrous consequences.

The 1986 Russell Street bombing, which killed a young constable and injured 22 people outside police headquarters in Melbourne's CBD, was a case of offenders with a pathological hatred of police; Family Court judges have been targeted over the years; and international terror groups use the internet to groom recruits who want to be heard.

What causes some people to develop a set against police, celebrities or the whole world? Where do fixations come from? And how do experts respond to people whose disappointment in life has hardened into a vicious grudge?

Who is prone to fixation?

There was a man who lived in the bush who would become angry at the world every now and again. The local cop knew that uniforms set him off and would always approach him dressed in casual clothes to calm him down. This time the local was on leave and the replacement turned up in a marked car. The angry man grabbed his gun and was eventually shot.

Some people who feel they have been wronged end up at newspaper offices as places of last resort. They have been to police, doctors, politicians and churches looking for relief. When they can't find the help they want, the conspiracy just gets bigger.

Since I started as a crime reporter in the late 1970s, I have had many such office visits. One day a lady turned up unannounced. She could not ring, she said, because her phone was tapped. She snuck out at night because police were watching her street and she could not speak to friends because those conversations were recorded. Another time, a distressed man said he would be punished for speaking the truth and would have gamma rays shot through his body by his enemies that night. Yet another arrived to ask why I had written a story about his life without permission. I explained I had never met him and had no idea what he was talking about. He started to shake violently, looked as though he was about to become physical and then ran away.

One man who revelled in his own anger spent years sending abusive emails because the newspaper would not publish a story that his entirely legitimate arrest was proof that he was the victim of police harassment. Another man reported a crime to police. He was not happy with the investigation and took it to federal authorities. During their investigations they found a small quantity of drugs and explosives at his property. He was charged and this led him on a 15-year campaign. Each time he wasn't believed he would conclude that authority was part of the conspiracy. Journalists wanted to write his story, he claimed, but corrupt editors would stop publication. One reporter was transferred overseas and another employed by the government to silence them. 'The media only write what is instigated by organised crime,' he claimed.

When he finally thought he was gaining traction, the Russell Street bomb exploded. 'Coincidence?' he asked. His conspiracy theory ultimately involved two prime ministers, senior police, judges, lawyers and journalists. He distributed more than 200,000 pamphlets, went on a hunger strike, protested in four states and had his claims examined by at least three public inquiries. He sacked lawyers who worked for him the moment they came back with the wrong answer. After a Supreme Court hearing lasting 29 days, his allegations were found to be baseless.

How do obsessions start?

In 2018, police established the $30 million Victoria Fixated Threat Assessment Centre, or VFTAC, designed as an early warning system – an elite group of specially trained police, forensic psychologists and mental health clinicians who find and stop potential lone attackers before they strike. It has quietly proven to be much more – the centre doesn't make headlines because its job is to stop the sort of events that do.

In 2006, the first such centre was established in Britain, primarily to identify stalker types who targeted public figures and celebrities. Since then, similar units have been set up not only in Victoria but in New South Wales, South Australia, Queensland

and Western Australia as well as in New Zealand and the Netherlands. While no one case sparked the initiative, counter-terrorism police have found that, in most single person attacks, offenders who are radicalised have a history of interaction with police and mental health services. The model of these centres is to break down the silos of knowledge, where medical and police authorities do not share important information.

> *In most single person attacks, offenders who are radicalised have a history of interaction with police and mental health services.*

In the case of the Lindt Café siege in Sydney's CBD in 2014, which resulted in the death of two hostages and the gunman, the offender had a criminal history, was considered violent, embraced radical political beliefs and had a known record of public obsessive behaviour, including chaining himself to a pole outside the New South Wales Parliament House.

Terror, it turns out, is not the main threat, and a strong law enforcement response can have dangerous consequences, feeding an aggrieved person's sense of injustice. According to Inspector Cooper, VFTAC 'prevents incidents by early identification of clients to disrupt the pathway that may lead them to harm others or themselves'.

Of the 300 cases that VFTAC had taken on by April 2021, three involved females. Most cases involving women are some sort of fabricated love interest with someone high profile, often a royal. These are considered sad but not dangerous.

At VFTAC, there are two types of clients. The first is fixated, where an offender blames an individual, usually someone in public office, such as a politician or judge, for their problems. The second (and 75 per cent of the cases) harbours a pathological grievance – where an institution such as police, courts or a company are blamed and 'need to be punished'.

In many of the cases, the individual has a point. It could be as small as a parking ticket or as big as a work dismissal, but the individual feels they have been wronged and denied justice. There is a moment when life derails. Something goes wrong, the obsessive fight to right the wrong leads to a drop in work performance, a demotion or dismissal, family issues and loss of friends. Instead of blaming the behaviour, the person blames the initial problem.

In almost every case of criminal mass destruction, there are threads discovered after the event that show what motivated the offender. The trick is to discover the human time bomb before it begins ticking.

How do you deal with someone who is fixated?

Since March 2018, VFTAC has examined 599 referrals (usually from medical staff, police and family) and has taken on 164 cases. At any one time, the centre is dealing with about 50 individual cases, each with its own dedicated detective, mental health expert and analyst.

The common complaint by street police is they cannot get mental health experts to a scene quickly enough to make a difference and, because of the need for medical confidentiality, may have to deal with a troubled offender without understanding existing conditions. At VFTAC, information is shared while making sure confidential medical information is not widely accessible. 'We have a prescribed sharing protocol and no health information is held on the police database,' says a senior clinical psychologist with the unit.

Cooper says where terrorism is involved, it is not the cause but the symptom and often the troubled individual is attracted to a conspiracy theory where governments and institutions can be blamed for real or imagined problems. Many terror suspects have underlying mental illness and have embraced radical and violent propaganda as 'the answer'. The worry is how quickly the desperate can be radicalised, which makes VFTAC a vital first defence system, using modern methods to follow the oldest rule in policing: it is better to stop a crime than solve one.

'Often when they have complained, they have been told to go away and that the problem doesn't exist. When we get involved it may be the first time they feel they have had someone listen,' says Cooper. 'The people here have to show empathy and not want to lock everyone up.'

Cooper says most cases require treatment rather than a traditional police response. In most cases, the team gives advice to the local police on how to handle the repeat, irrational offender. Those cases that do end up at VFTAC, involving people with the most intense pathological grievances, can be leading desperately lonely lives, often self-imprisoned in boarded-up houses believing they are the victims of monstrous conspiracies. VFTAC tries to reintegrate their clients into society 'before they harm themselves or others'.

Take Frank*, 43, who alleges he was assaulted by police several years earlier during an arrest. His claims were denied by the police involved and, without any corroborating evidence, the case was dismissed. Frank is a prodigious letter writer and for years sent written complaints to the 'offending' police station demanding compensation and an apology. With every rejection he became angrier, shifting his attention to a state cabinet minister.

When the minister couldn't help, he became part of the conspiracy and was branded corrupt like the rest of them. Frank stopped getting mental treatment, withdrew from his few remaining associates and his letters went from angry to menacing.

---

* *Hybrid cases from VFTAC files*

When the centre became involved, they encouraged him back to mental health treatment, helped him get legal aid at the Family Court and linked him to a community policing unit that saw him as a client rather than a troublesome crook. The aim was to release the overwhelming pressure Frank felt was destroying his life. Finally, he felt someone was listening and his fixation on the minister lapsed.

Another case involved 25-year-old Harry*, who was no longer eligible for WorkCover payments following an updated medical report. The insurance company was adamant. Harry was also adamant. He was sick of being ignored and being 'treated as a number'. They were stealing his money, he claimed, and he would make them pay. If they continued to ignore him, he would kill himself inside their office.

He had no history of mental illness but had been pushed to the edge by his injuries and what he thought was his non-existent future. The insurance company referred the case to police and VFTAC. A team linked with Harry brought his family into his safety net and helped him engage in alcohol and drug treatment. It also persuaded the insurance company to help provide paid mental health support while he underwent vocational training to get back on his feet. His grievance subsided.

How kind can you be?

The man with the Queensland drawl was anything but laconic. Having moved to country Victoria, Ian* engineered confrontations with local police and claimed to want to commit 'suicide by cop'. His VFTAC file stated: 'The referred person shows evidence of extreme grievance, hatred, bias, perceived injustice and/or resentment which is highly personalised or idiosyncratic, rather than a widely shared ideology.'

In other words, he hated cops and didn't know why.

In Queensland, he was obsessed with an ex-partner and, when charged with stalking, his obsession turned to the police. Moving to Victoria for a new start, he took his hatred with him. At first, police thought he was just a knucklehead, the sort who had to have the last word and was never wrong. But his behaviour became increasingly erratic and worrying.

In court, while he was reporting for bail in Victoria over Queensland offences, he became agitated and started ranting that if he killed the magistrate who had granted an order against him 'he would be taken seriously'. He said Queensland police 'were trying to take my life away' and if 'I drive through a police station I'd get some attention'.

Afterwards, he started to show a concerning interest in the female police involved in his case, repeatedly ringing the station to try to talk to any of them.

When VFTAC took the case, they made sure Ian received mental health treatment, notified Queensland police of the threat to the magistrate, offered to take out protection orders against him, checked he had no access to firearms, set up an older male police officer as his case manager, consolidated court dates to lower his stress and anger levels – and tried to break down his pathological hatreds. They have put danger flags on his file so any police who deal with him know he has the capacity to become aggressive.

They used the carrot and stick approach, telling him they would do all they could to help him if he behaved. But they also told him there was a big stick they could use if he continued down his present path.

Sometimes all the love in the world is not enough.

11

WHAT MAKES AUSTRALIA'S WILDLIFE UNIQUE?

You'll not find a wombat, numbat or truffle-snuffling potoroo in the wild outside of Australia. In fact, elsewhere in the world there is nothing remotely related to many of our plants and animals. So what does it mean when a species becomes extinct?

Mike Foley

Australia's wildlife was so weird to the nineteenth-century scientists of the northern hemisphere it caused one of them, a devout Christian by the name of Charles Darwin, to question God's creation. 'A disbeliever in everything beyond his own reason might exclaim, "Surely two distinct Creators must have been [at] work; their object however has been the same,"' wrote Darwin in 1836, after visiting the Coxs River at Wallerawang in New South Wales. A day's kangaroo hunting, during which he had stumbled on a potoroo, parrots and a platypus, had blown Darwin's mind.

He hadn't known it then but most of the plants and animals he had encountered had spent 40 million years evolving their own unique solutions to a common conundrum – how to survive and prosper. The longer Australia's plants and animals spent on an island in glorious isolation – and 40 million years is a decent whack of time even in the slow game of evolution – the further its species reproduced along their own evolutionary branch, so that their solutions to these issues were, compared to the rest of the world, odd.

More than 80 per cent of Australia's nearly 400 mammal species, from flying marsupial sugar gliders to truffle-eating miniature kangaroos, are found nowhere else in the world. The same goes for about 95 per cent of nearly 1000 reptile species, from blue tongue lizards to thorny devils, and for more than 90 per cent of the nearly 300,000 species of invertebrates, of which fewer than 15 per cent have been formally described.

'It's not just the sheer numbers of species that matter in Australia,' says the Wilderness Society's Tim Beshara, 'it's how totally unrelated these species are to everything else on the planet. Australia's flora and fauna species don't just represent a few leaves on the tree of life, they represent entire branches and some of the trunk.' This means that when just one species becomes extinct in Australia, a significant part of Earth's evolutionary history is lost – there is no near cousin, nothing quite like it, to fill its niche.

Our plants and animals are, says Beshara, 'disproportionately special'.

> *Australia's flora and fauna species don't just represent a few leaves on the tree of life, they represent entire branches and some of the trunk.*

So how did our flora and fauna get to be so unique? What's so unusual about creatures such as kookaburras, lyrebirds and thorny devils? And what are the consequences if they become extinct?

How did our plants and animals end up on their own?

The Gondwana supercontinent, comprising Africa, South America, India, Madagascar, Australia and New Zealand, began to slowly break apart 165 million years ago. Australia and South America remained linked to Antarctica by land bridges covered in beech forest over which marsupials were able to roam. Australia drifted away from Antarctica about 40 million years ago, which severed the link to South America as well.

Further north, Australia and New Guinea, which sit on the same tectonic plate, were drifting northwards as they still do today: the Australia–New Guinea continental plate is colliding with Eurasia, forcing up the New Guinea Highlands and buckling downwards into a now-submerged land bridge across the Torres Strait. That land bridge sank just 12,000 years ago, at the end of the last Ice Age, when melting glaciers filled in the trough – which is why there are common species among Australia, Papua New Guinea and the Indonesian provinces to its near west, notably the cassowary as well as many birds, echidnas and tree kangaroos.

Fossil records reveal Australia's ancient ecological isolation. 'All of the other continents,' says eminent Australian paleontologist

Michael Archer, 'had physical connections between each of them, meaning that they have been exchanging biota [animals and plants] back and forth.

'We have, for example, camels in South America and camels in Asia. The camels in South America are the llamas and alpacas. South America has had elephants and lions, all the same things that were spread all the way across the other continents.

'Essentially, the other continents have had a shared genome for much of the last 65 million years, since that meteor hit the Earth and obliterated the non-flying dinosaurs, when there was a massive explosion in the northern hemisphere of placental mammals.'

What kinds of animals are unique in Australia?

In 1859, British naturalist Alfred Russel Wallace drew a line on a map through the Indonesian archipelago representing his realisation of a geographic split in mammal evolution. In the west were placental mammals, nourished by their mother's uterus, and in the east were marsupials that suckle their young in the pouch, and monotremes that lay eggs. The great gulf in evolution is cut sharply on the map. On the ground, from the northern tip of Lombok on the marsupial side of the Wallace Line, the lands of placental mammals can be glimpsed across a narrow strait, just 35 kilometres away in Bali. Wallace's contemporary, Thomas Huxley, altered the northern tip of the line to incorporate the Philippines, which has a lot of endemic wildlife.

Where deer evolved in the northern hemisphere, kangaroos cropped up in Australia. Primates evolved in Africa and migrated into South and Central America as well as Asia, many growing limbs to exploit fruiting trees. But in Australia marsupials fill that role, with tree kangaroos and cuscuses that swing from limb to limb. And even more ancient animal lineages remain within the echidnas and platypuses, monotremes that uniquely among land animals use electrolocation to sense prey and even retain the reptilian traits of egg-laying and legs on the sides of their bodies rather than underneath as with mammals.

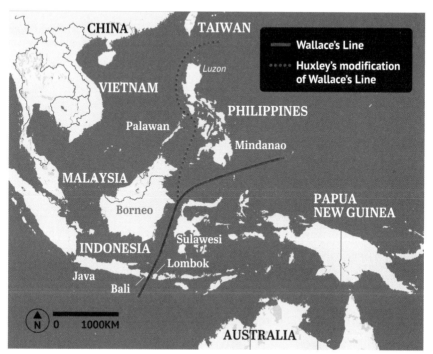

Map by Jamie Brown

Pioneering ecologist Professor Christopher Dickman says marsupials arrived in Australia via the land bridge from South America about 40 million years ago, and there was a 'spectacular radiation', meaning the animals got busy evolving into a wide array of species. 'Because of the integrity of the radiation, we've got all of these forms that really don't occur anywhere else. Certainly, there are marsupials in South America but they're all little guys. They're all five kilograms, at the very maximum,' Dickman says. 'The vast majority of them are smaller, mostly omnivorous and they're pretty similar to each other in many respects, whereas in Australia they've been able to radiate into the country's many environments.'

A marsupial lion weighing about 130 kilograms roamed Australia's forests, the largest ever carnivorous marsupial before it died out 30,000 years ago – about 20,000 years after the arrival of Aboriginal Australians.

The next biggest Australian meat-eater that cohabited with humans was the thylacine, or Tasmanian tiger, its males weighing in at 20 kilograms; and the next biggest after that was the Tasmanian devil.

And it wasn't just the carnivorous marsupials that proliferated.

'We've got specialist insect-eating marsupials, specialist termite eaters – which is primarily the numbat – and we've got the world's only specialist fungivore marsupials in the potoroos and bettongs.'

West of the Wallace Line the aardvark evolved as the mammalian solution to exploit the food resource of protein-rich termites, but in Australia, in a process known as convergent evolution, the marsupial solution to fill this niche was the numbat. Numbats are unique among marsupials as the only species active in the day. They once lived across vast tracts of Australia but have been hunted down by foxes to a small corner of Western Australia.

Another case of convergent evolution is the tree kangaroo. In the absence of competition from arboreal primates such as monkeys, lemurs and gibbons, kangaroos adapted to climb trees and eat the fruits and leaves on offer.

Potoroos, which so amazed Darwin, as did their relatives the bettongs, are unusual not just because they're the only marsupials that dine on fungi – native truffles buried in the soil, to be precise – but also for their role as ecosystem engineers. 'In the forest environment, potoroos eat fungi and they move the spores around in the forest,' Dickman says. 'They're the vectors for the spores.'

Fungi have a symbiotic relationship with some plants, where they source necessary carbohydrates from plant roots and in return harvest scarce nutrients from the soil such as phosphorus, nitrogen and potassium, which dramatically improve plant growth. 'You lose the potoroos and you lose the ability for the fungi to move around in the forest,' says Dickman. 'So in the wake of a disturbance, like a fire or a flood or somebody goes through with the bulldozer, if the fungi is not there, and the potoroos have gone, plant growth will be much, much slower.'

Did the world's songbirds really come from Australia?

When Darwin sat down to write after his trip to Coxs River, it was little wonder he questioned how one god could create both those European animals he was familiar with and those shocking creatures he'd witnessed on the western side of the Blue Mountains. 'Would any two workmen ever hit on so beautiful, so simple and yet so artificial a contrivance? I cannot think so,' Darwin wrote. Many scientists and historians argue this moment was the genesis of Darwin's revolutionary theory of evolution, which was eventually published in *On the Origin of Species* in 1859.

Even the most mobile of animals, the birds, boast species in the hundreds that are endemic – that is, unique – to Australia. Of the 830 types of bird in Australia, 45 per cent are found only here. What's more, as biologist Tim Low points out in his outstanding 2017 book *Where Song Began*, all the world's songbirds and the most visually appealing birds have Australian roots. 'An Australian origin is implied for every songster in an English country garden, for all the chickadees, cardinals and jays in America, for bulbuls, babblers and sunbirds in Asia, and weavers, whydahs and bush-shrikes in Africa,' Low says.

One of Australia's most spectacular songbirds, the superb lyrebird, also has the deepest evolutionary lineage and retains physical features of the earliest songbirds. The lyrebird's mimicry is not limited to just about every other bird in the forest but extends to human voices, camera shutters, mobile phones, car horns, even chainsaws. It is also one of the world's biggest songbirds, with a dramatic fantail used in courtship displays. So well adapted are the largely solitary lyrebirds to their forest environments, they have been known to anticipate incoming bushfires by congregating in sheltered gullies and watercourses to survive the flames.

It's now known songbirds spread out around the globe from Down Under but this fact was only scientifically recognised in 2004, due to the parochial assumptions of the majority of the world's northern hemisphere–based ecologists. 'Going back a couple of decades – maybe it was because of cultural cringe – we always assumed that songbirds must have arrived in Australia and then radiated here. But now, we realised that actually the radiation took place here,' says ANU professor of wildlife ecology Sarah Legge.

Among the most captivating of these 'weird' birds, she nominates the palm cockatoo and the eclectus parrot, both found in the forests of northern Australia and New Guinea. 'Palm cockatoos are the only bird that uses a tool to make music. They get a stick, which they often prune a bit, and they use that to bash the side of their nesting hollows in a rhythm,' Legge says. 'That's way out there, that's really unusual. What's really incredible about the eclectus parrot is they've got that reversed sexual dichromatism – so the females are bright red and the males are green so they're more camouflaged. But then the females are also promiscuous, so they'll mate with multiple males and then sit in the nest hollow for months at a time getting fed by multiple males – they've got a really wacky lifestyle.'

Weird bird behaviour isn't limited to Australia's remote forests – witness the blue-winged kookaburra. 'The incredible thing about them is that, as adults, they're in cooperatives. Kookaburras live in family groups where the young from previous nestings stay with their parents and help them raise more broods in following years. But when they hatch in the nest, because usually two or three chicks hatch in the nest, they're siblicidal – so the chicks, within minutes of hatching try to kill each other,' Legge says.

'In this one species, you've got this most extreme selfish behaviour, I guess at one end of their life. Killing siblings is a big deal because they share genes. And then, at the other end of their life, they forgo their own breeding attempts to help feed and care for their parents' chicks. They flip from one extreme to the other.'

Palm cockatoo by Matt Davidson

So why are we losing species?

The extraordinary uniqueness of Australia's natural endowment is not well understood, nor is it adequately protected by environment laws, says the Wilderness Society's Tim Beshara. 'There is an under-appreciation among politicians, business leaders, and even the wider conservation movement, about how disproportionately special Australia's biodiversity is,' he says. 'None of the government regulation or industry standards take into account Australia's special status among nations.

'For instance, platypus and echidnas make up one of the three major branches of the mammal family tree, marsupials make up the second branch, and all of the rest of the cats and bears and whales and rats make up the third and final branch. That means we are the centre of the world for two out of three major groups of mammals.

'To lose a species in Australia means there is a much higher likelihood of losing entire swathes of evolutionary history than a species lost anywhere else. As the world's most evolutionarily distinct realm on the planet we have a higher obligation to protect what we have – and, currently, we are failing to do so.' The thylacine was finally hunted to extinction in 1936, having died out on the mainland 3000 years ago along with the devil, possibly due to competition with dingoes, which were introduced between 5000 and 10,000 years ago.

There is an under-appreciation among politicians, business leaders, and even the wider conservation movement, about how disproportionately special Australia's biodiversity is.

Australia has lost 39 species of mammals since colonisation, with the nation's share of the world's extinct mammals totalling 38 per cent. The growth in farming and feral cat numbers, for example, is thought to have wiped out the diminutive Eastern hare-wallaby, which lived in the grasslands of south-eastern Australia, and the crescent nail-tail wallaby, native to central and south-west Australia.

All up, about 100 of Australia's native flora and fauna species, from rare orchids to the Daintree's river banana, have been wiped off the planet. Four species of frogs and 22 species of birds, including the paradise parrot and the red-crowned parakeet, have vanished forever. Many more creatures are critically endangered. The rate of loss, among the highest in the world, has not slowed over the past 200 years, since the early days of land clearing and the introduction of feral cats and foxes.

University of Sydney professor of ecology and evolution Glenda Wardle says a shift in thinking about environmental conservation is underway, guided by the principle 'that all species have a right to exist'. 'We don't just need to keep the ones that are good for giving us medicine, or the ones that give us food and all the rest is somehow dispensable – and maybe put a few in a reserve somewhere. The idea now is that all species have a right to life,' she says.

Ring-fencing reserves for animals in out-of-the-way places not needed for farming, building or mining will not be enough to conserve Australia's biodiversity, especially in the face of human-induced climate change, Wardle says. 'Species are evolving, always. It's not a "turn it on, turn it off" thing. They're always matching to what the conditions are. And as we're going through this rapid climate change we expect more rapid change in some species, in terms of how their traits match the environment.' Research indicates that Australian birds from temperate zones have adapted to higher temperatures with decreasing body size.

Can we recover what is lost?

Tens of millions of years of this unique evolutionary history are carried in the genes of animals, and scientists are just beginning to glimpse ways to unlock the hidden treasure. Advances in science mean we can recover ancient genomes and so, potentially, lost species. The technology is still developing and will require ethical and environmental consideration. 'No other continent can claim a similarly unique library of genetic information in the way that Australia can,' Archer says, arguing that Australians have a 'greater responsibility for securing the world's genome than anywhere else in the world'.

*Advances in science mean we can recover
ancient genomes and so, potentially, lost species.*

The rate of loss, among the highest in the world, has not slowed over the past 200 years.

For example, Archer says the genetic information, or genome, of koalas and wombats – which are one another's closest living relatives – represent a staggeringly diverse range of marsupial species that once walked the Earth, including the largest marsupial ever, the giant wombat-like diprotodon optatum and Australia's own marsupial lion. 'All of these genomes are probably represented, at least in significant part, within the living genomes of the koala and the wombat. So, they're more than what they represent in their own right – because every animal on Earth has a genome that has a staggeringly huge amount of silent genes that relate to its ancestry. And the more we understand about gene manipulation – and that's happening very fast – the ability to reach into the living animals' genomes and recover ancient genomes becomes very, very exciting.

'We begin to keep our fingers crossed that what we've lost may not have to be lost forever.'

Archer has started a project to try to recover the Tasmanian tiger – 'and what we've discovered is we now have the whole genome, all the information, we technically need to do that'.

There are other good reasons to preserve our existing species, too. 'Koalas have a special antibiotic in their pouch that helps to keep their youngsters free of diseases but it's not a class of antibiotics we've ever seen, or that we're used to using,' says Archer. 'So, reaching into the koala, in effect, to find out what is this antibiotic that it's making – and what are the genes that are doing it – could be extremely important for us.'

Australia has been a continent-sized sanctuary for millennia but for the past 230 years, species have disappeared quickly and consistently. Nascent genetic technology to recover lost species is exciting, but it won't fix a key cause of extinctions – habitat loss. Political will is required to drive reforms of our laws on the environment and planning.

12

WHAT IS CONCUSSION AND WHAT IS CTE?

Once, head knocks were part of football.
Now there are rules governing them.
How much harm can a jolt do?

Konrad Marshall

The AFL world was stunned by the untimely death of legendary St Kilda defender Danny Frawley in 2019. Perhaps the greater shock came later, however, when Frawley was found to have been suffering from CTE, the degenerative brain disease often associated with athletes who have a history of head knocks and concussion. Frawley was only the second former AFL player to be diagnosed with CTE, following its discovery in the brain of the late Graham 'Polly' Farmer in 2020, establishing an incremental link between head knocks on the field and tragic mental health problems later in life.

In the AFL, roughly 60 concussions are recorded every season, and every one creates its own reverberation of commentary and dissection. The chatter spreads, and seeps into the minds of administrators, and parents, and custodians of our football codes at all levels, prompting an inevitable question: is football even safe any more?

Yet the concussion conundrum is far from simple. Experts in neuroscience and sports law, club doctors and former players, administrators and union officials, seldom find much shared ground on this medico-legal complexity. 'Uncertainty is driving this vast difference of opinion,' says neuroscientist Dr Alan Pearce, from La Trobe University. 'There's so much we don't know – so many questions we have to answer.'

What are they?

What happens to the brain?

Concussion is a transient injury caused by any jolt to the head (or body) that delivers an impulsive shock to the brain, causing it to rock back and forth in the skull or twist on its axis. It's a functional neurological disturbance rather than a structural injury, and it hides itself well. It will not show up conclusively under X-ray or CT scan or MRI, or in tests of blood and saliva.

In fact, concussion is diagnosed only by observing overt symptoms such as dizziness and confusion, nausea and unsteadiness. Complicating matters, the condition varies wildly between individuals and incidents. Symptoms can manifest instantaneously – or appear hours later. And they can linger for months – or disappear within minutes.

Chronic Traumatic Encephalopathy (CTE) is a progressive brain disease found in people with a long history of head trauma – not so much a handful of big concussions but rather hundreds (or thousands) of smaller impacts over a number of years.

> *CTE often manifests as a kind of dementia but can only be diagnosed post-mortem.*

The Concussion Legacy Foundation, the advocacy arm of the groundbreaking concussion research group at Boston University, makes a distinction between bigger hits and 'subconcussive impacts' (of the kind sustained when NFL players routinely bang their helmets into one another). The foundation uses the analogy of a car driving down a poorly maintained road. Sure, big potholes might burst a tyre or crack an axle but smaller potholes do immense harm, too: drive over those little bumps a dozen times a day, every day of the year, for more than a decade, and the wear and tear can be catastrophic on the car.

It makes sense, then, that CTE was first diagnosed in 1928 in boxers, under the descriptor 'dementia pugilistica' (commonly known as 'punch drunk syndrome' and later 'slug nutty') but it roared back into the public consciousness in 2005, when pathologist Bennet Omalu found CTE in the brain of an American footballer, former Pittsburgh Steeler Mike Webster. This discovery was subsequently made into a movie starring Will Smith as Dr Omalu (although an investigation by *The Washington Post*, 'From Scientist to Salesman', calls into question some of Omalu's claims).

CTE often manifests as a kind of dementia but can only be diagnosed post-mortem. Once a person who has pledged their brain dies, they are sent to a mortuary where their brain is removed and weighed then fixed in formalin to preserve the tissue. Researchers scrutinise photos of the brain for bleeding, bruising and patterns of atrophy, where the sulci (valleys) and gyri (bumps) of the brain have become deep or shallow, narrow or wide.

They then slice the brain vertically at three-millimetre intervals along what's called the coronal plane, photograph each slice then place portions the size of a postage stamp into microscope slides. The slides are washed in various stains that react to proteins and – in the search for CTE – scientists look specifically for the dark brown build-up of a protein known as tau. 'Tau normally helps a healthy neuron,' says Pearce. 'But if an accumulation of it clumps, it can strangulate a neuron or nerve cell. Found in specific places, it becomes evidence of a degenerated brain.'

When has CTE been found in Australian athletes?

The first diagnosed case was former rugby union player Barry 'Tizza' Taylor, who died in 2014, aged 77. His brain was sent to the Global Brain Bank at Boston University and displayed all the hallmarks of severe CTE. Soon after, the local arm, the Australian Sports Brain Bank, was launched as a collaboration between Sydney University and Royal Prince Alfred Hospital, and dozens of retired athletes pledged to donate their brains upon death.

In 2019, rugby league was rocked when the brains of former NRL players Steve Folkes and Peter Moscatt were examined and both found to have the condition. That same year, however, the first donated brain of a former Australian Rules player, WAFL player Ross Grlusich, was examined and, despite Grlusich suffering numerous concussions throughout his career and ultimately dying of dementia, no evidence of CTE was found. (As one researcher has noted: 'Not all smokers get lung cancer.')

In February 2020, CTE was officially diagnosed in a former AFL player – the legendary player and coach Graham 'Polly' Farmer, who died in late 2019 aged 84 after battling dementia for a number of years. Next, it was found in the brain of Saints champion Frawley, whose battles with mental health were well known. In early 2021, it was revealed that former Richmond midfielder Shane Tuck – who ended his own life in 2020 at 38, while dealing with mental health issues – suffered from severe CTE, which one neuropathologist described as 'the worst case I've seen so far'. Tuck's diagnosis was particularly stark, as the 173-game Tiger veteran had no reported history of serious concussions.

What long-term problems can head knocks cause?

A number of scientific papers have found a correlation – as distinct from a causal relationship – between multiple concussions (three or more) and a greater chance of cognitive impairment later in life, increasing the risk of everything from anxiety to epilepsy, Parkinson's disease and the nervous system disease ALS, also known as Lou Gehrig's disease. People who sustain even mild traumatic brain injuries often experience underlying neurological problems at an accelerated rate, notes Dr Mark Cook, chair of medicine at the University of Melbourne and director of neurology at St Vincent's Hospital.

'The most famous example of repeated brain injury causing problems down the track was Muhammad Ali [who was diagnosed with Parkinson's disease following his storied boxing career]. Nothing was obviously wrong right away but down the track everything was wrong,' Cook says. 'I like boxing and I like football and I like ice-hockey, but if you injure your brain repeatedly, it's hardly surprising that it could lead to seizures and cognitive decline. Excuse the pun, it's a no-brainer.'

There has only been one longitudinal study of head knocks in Australian athletes, by Dr David Maddocks of the University of Melbourne, who tracked concussions in AFL players at three clubs in 1989. A quarter of a century later, in 2014, University of Melbourne masters student Hannah Blaine (with help from Maddocks and supervision from Professor Michael Saling) re-tested players from one of the clubs. 'We found that concussion didn't have an impact on their cognitive or psychosocial functioning,' Blaine says. 'One guy had reported 20 concussions. He actually performed the best across the sample. Particularly on things like learning and memory.' They presented the findings at the fifth International Neuropsychological Society and Australasian Society for the Study of Brain Impairment conference in Sydney in 2015. But their message – that this cohort of footballers was unaffected by concussion later in life – was unwelcome.

Concussion is a notoriously difficult condition to study, and occasionally throws up results such as those discovered by Blaine, yet it should in no way dispel any urgency to address the issue, given the raft of young players retiring after enduring concussions too severe or too frequent. The list grows every AFL season, to more than a dozen in the past few years, including the likes of Koby Stevens, Kade Kolodjashnij and Liam Picken. Others remain in limbo after a string of hits, including Patrick McCartin, 25, the number one pick from the 2014 draft, who said in an interview in 2019, 'I'm a shell of a person that I was, really. I'm completely different.' (McCartin played for the Sydney Swans reserves in 2021 in the hopes of resurrecting his career after more than three years out of the game.)

NRL, too, has been affected. Newcastle winger James McManus was forced out of the game after enduring a series of head knocks, while former Rooster Eloni Vunakece admitted a string of heavy blows played a part in his decision to quit the sport. Sydney Roosters co-captain Jake Friend – a three-time premiership winner – retired after receiving three concussions in six months, following more than 20 concussions throughout his career.

And rugby union has not gone untouched, with former Wallaby players Toby Smith and Anthony Fainga'a leaving the game for fear of what might happen should they suffer another concussion.

Then there are retired players of yesteryear in each code who are suffering debilitating cognitive issues, including memory loss and mood swings, confusion and seizures. Many have already pledged to donate their brains to the Australian Sports Brain Bank. Player agent Peter Jess has been in touch with athletes from several sports, and says there is a consistency to the issues they report.

'Depression. Anxiety. They can't find their cars, they can't find their car keys – along with erratic behaviour, and their wives about to leave them,' says Jess. 'That's the elephant in the room – a lot of these guys are on wife two and three. And they'll tell you, "That guy over there is not the guy I married. He's short-tempered. He's angry. He can't hold down a job." It's a really sad situation.'

So what happens on the field now when a player is hit?
All codes have different protocols but they all use the diagnostic SCAT-5 (Sport Concussion Assessment Tool, Fifth Edition). First, a club doctor approaches the player and asks questions such as 'Who are you playing on?', 'What's the score?' and 'What day is it?' These are known as 'the Maddocks questions' because they were devised by Dr Maddocks in 1995 after his seminal paper, 'Neuropsychological recovery after concussion in Australian rules footballers'. Depending on the quality of the player's answers – and red flags such as stumbling and disorientation – the player comes off the ground for 20 minutes. They are checked for obvious physical symptoms such as double vision, vomiting, convulsions – or headache, fatigue, nervousness.

A cognitive screening then tests memory and recall. The doctor might read aloud a series of words – finger, penny, blanket, lemon, insect – and ask the player to repeat the string of words, in any order. The doctor does the same with digits, the list of numbers growing progressively harder to remember. They also perform a balance examination (walking a line on the floor, heel to toe, nine metres long). All of this contributes to a subjective read on their state of mind. If they are deemed to have been concussed, they are removed from play.

Every concussion is unique, says Dr Robert Cantu, co-founder of the CTE Centre at Boston University, senior adviser to the NFL Head, Neck and Spine Committee and the world's foremost expert in 'return to play' procedures.

Someone who has suffered another concussion only recently, for instance, might need longer to recover than someone concussed for the first time. Someone who has been concussed many times will probably need to rest even longer again. 'And most importantly,' says Cantu, 'someone who's had a previous concussion whose symptoms have lasted weeks or months should certainly be given more time off than someone whose previous concussion cleared up in a matter of minutes.'

An understanding of what happened in the incident is important, too. Was the collision horrific, where you would expect or predict concussion? Or was the blow fairly inconsequential or innocuous? The latter is actually more troubling because it took such little impact to cause harm. 'If the hit seemed minor, you're going to be more cautious putting that person back into play because obviously that person is going to be exposed to other such minor hits.'

What's the risk in returning early?

There are a few dangers to consider. First, if a player hasn't completely recovered from concussion, their athletic and evasive skills could be compromised. (Symptoms often vanish quickly but other subtle changes in the brain are more persistent.) Professor Michael Saling was one of the co-authors of the first Australian studies into concussion in 1989, which found persistent impairment in speed of information processing, reaction time and decision-making. 'We could see that the cerebral effects of concussion were quite long-lasting, and the cycle of recovery could extend over a two-week or three-week period.'

Second, if a player who has not fully recovered from concussion sustains a follow-up head trauma, their symptoms can be exacerbated. 'It can be debilitating,' says Cantu. 'Someone who might have been days from overcoming their concussion may end up with a concussion that is prolonged weeks, or months, or more.'

Third, if a player sustains one of these hits while not fully recovered, they may be more susceptible to cognitive decline later in life. Associate Professor Sandy Shultz of Monash University and the Alfred Centre has been studying this idea of a 'window of vulnerability' since 2010. His work involves experimenting with 'pre-clinical models', meaning rats, which wear tiny 3D-printed helmets and are then concussed with a mechanical device. When these rats were struck again – while biological imbalances from the first concussion were still in play – they experienced progressive and persisting long-term brain damage, as well as learning and memory deficits, sensorimotor abnormalities, depression and anxiety.

But what this means for footballers is unclear. 'Because how do you study that in people?' asks Shultz. (You can't take a statistically significant group of healthy young athletes, hit them over the head, then hit them again at varying intervals during their recovery, just to see what happens.) The data from rodents suggests the timing of subsequent hits is important, says Shultz, 'but whether that applies in humans, we just don't know'.

Until recently, roughly four out of five players who suffered a concussion while playing AFL still lined up to play the next weekend. That might sound an alarming number, but perhaps not when considering the research of Dr Nathan Gibbs, a former NRL player, club doctor for the Sydney Swans for nearly two decades and now head doctor for the Wallabies. During one 12-year stretch at the Swans, Gibbs compiled his own research with surprising results. In the aftermath of 140 concussions, every single Swan played the following week. And their immediate performance – based on a mark out of 20 given by the coach – was unaffected. 'They played well,' Gibbs says. 'The outcomes were good.'

> *If players know that being concussed will rule them out of a game the following week, it might motivate them to conceal their symptoms.*

Still, the AFL moved to tighten its concussion protocols in 2020, requiring players to successfully pass the SCAT-5 (the same test used during games to diagnose concussion) a full five days before playing again, instead of merely a day in advance of a game, meaning a player would essentially need to pass the test within a few days of their concussion, or miss the following match. This seemed like a careful step in the right direction, given concerns that a blanket rule simply sidelining every concussed player for a week would be difficult to implement. The persistent fear – in all sports – was that such a rule could force concussion 'underground'. That is, if players know that being concussed will rule them out of a game the following week, it might motivate them to conceal their symptoms.

This assertion is not without merit. Players have recently admitted to withholding the truth about their symptoms from club doctors, while retired players have admitted to intentionally performing poorly on their mandatory pre-season baseline cognitive testing, making it easier to pass a concussion test when hit during the season.

And so in 2021 the AFL took their cautious stance on the issue further than any code, bolstering their protocols with a new rule: if a player is medically diagnosed as suffering a concussion, they are now automatically sidelined until the twelfth day after the day on which they were concussed.

Pundits immediately asked, 'What if that causes a player to miss a grand final?' and almost immediately received an answer, when AFLW star Chelsea Randall was concussed in the women's league preliminary final, and ruled out of the 2021 AFLW grand final one week later. The Adelaide Crows captain was asked at first whether she would challenge the new protocols, in an attempt to take the field. 'I decided not to take any further action because what kind of message would that be sending to our grassroots football?' she told her club media team. 'Because concussion is serious, it is scary.'

However, former Saints skipper Nick Riewoldt, who played in two losing grand finals and a draw in his 336-game career with St Kilda, told Fox Footy he would consider legal action if faced with the same predicament as Randall. Riewoldt is adamant that the fallout from concussion be taken seriously, yet concedes the drive to win a premiership is powerful. 'If I was in the same situation, imagine September, still playing in your 30s, captain, all of those things. I'm taking it as far as I can take it. I'm going to the Supreme Court, I'm going for an injunction,' Riewoldt told Fox Footy.

Although Riewoldt's comments were criticised, they reflected the sentiments of a number of players. The AFL has admitted it is open to discussing the prospect of a pre-grand-final bye beyond 2021 that might give players concussed in a preliminary final the recovery time to play in a grand final.

Are retired players likely to sue for compensation?
A class action in the US brought by retired gridiron players against the NFL, and settled in 2013, has resulted in more than

US$600 million in claims so far, and could balloon well beyond that amount. Naturally, people have considered similar lawsuits in Australian sport.

A handful of former AFL players – including John Platten, John Barnes and Shaun Smith – have investigated pursuing a class action lawsuit, proposed by Adelaide lawyer Greg Griffin.

Two Sydney law firms, Bannister Law and Cahill Lawyers, are investigating similar suits for former NRL players suffering 'reasonably preventable brain injuries'.

These actions would allege that the NRL and AFL – as the sole controller of rules, medical panels, protocols and sanctions – allowed their games to be needlessly violent in various ways, such as routinely allowing players back into the cauldron of training or games when they were demonstrably unwell.

However, the suits will face some challenges. First, class action lawsuits are incredibly expensive, and so will require the investment of a litigation funder. Next, to be certified as a class, they will need to show that the players are all facing similar issues as a result of similar harm caused by similar negligence. (That could be tough, given that players come from different eras, and might have even received some of their concussions in state or amateur leagues, in childhood, or even post-career sports like boxing.) They will also need to establish 'causation' – proving in court that the maladies they now endure are a direct result of the blows they sustained on the field, which is difficult given CTE cannot be diagnosed while still alive. And they will need to show a negligent breach of a duty of care, establishing exactly when and what the AFL and NRL knew about concussion, or should have known, along with their failure to act.

The case mooted against the AFL was raised three years ago but no statement of claim has been filed in court.

Smith received a $1.4 million insurance payout from MLC Life Insurance in 2020 after he was found to have suffered 'total and permanent disablement' due to head knocks he copped in his 109-game AFL career.

Individual players can take legal action, too, arguing they were injured during a period in which leagues or clubs knew enough but didn't act. In the NRL, for example, former Newcastle and NSW winger James McManus has attempted to sue the Knights over their management of his head knocks, claiming they failed to properly assess or monitor him, and continually exposed him to new danger after suffering concussions. In the AFL, former Adelaide Crow Sam Shaw has done the same – believing the concussions that ultimately forced him out of football were mishandled by Crows medical staff.

A more likely path to restitution would be some kind of fund, established to support any player whose livelihood has been diminished by the debilitating effects of concussion. (Athletes, unlike other employees in Australia, have no workers compensation system to rely on for 'no fault' benefits when hurt performing their profession.) The AFL has so far resisted calls to establish a concussion-specific fund (as the NFL has done in America), but has instead suggested it will put more money into the AFL Players' Association Hardship Fund, which supports players who are struggling in post-playing life.

How seriously are football codes taking the issue?

Very seriously. It's worth noting that players such as Frawley and Farmer took to the field in a wildly different era, when little was known about the potential fallout from head knocks. Once the extent of the insidious damage wrought by concussion in the NFL started to emerge in the mid-2000s, local leagues began to act. In 2008, for instance, the AFL codified the way players should be assessed after head knocks, and in 2009 and 2010 – and a handful of times since then – they tweaked the laws of the game to curtail reckless high bumps and dangerous tackles, also moving to disincentivise players who try to draw free kicks for head-high contact.

Rugby union made strides, too, lowering the legal tackle height worldwide (from shoulder to armpit) in 2019, which led to an instant 28 per cent drop in concussions. They also introduced the Blue Card system, under which medical staff or referees can remove a player from a game the moment they see any symptoms of concussion. In the junior ranks, rugby created the Size for Age program, in which juniors can be moved between age groups to suit their physical or mental development.

For young rugby league players, there is now a modified rules SafePlay format of the game, along with the TackleSafe program to teach juniors better tackling technique. The NRL, meanwhile, began doling out huge fines to clubs that allowed players to remain on the field after a concussion. The Wests Tigers, Canterbury-Bankstown Bulldogs and the Parramatta Eels are among many on the receiving end of $20,000 sanctions.

Sideline technology has improved, too, allowing medical staff in all codes to replay incidents using the Hawk Eye tablet technology. In 2019, the AFL also began employing independent 'spotters' to identify potential concussion events.

Helmets can only go so far in offering protection. AFL and NRL and rugby union players can (and some do) wear soft rubber and foam helmets, but these are more likely to prevent bruises and cuts and potential fractures than concussions. Bear in mind that players can sustain a concussion when no contact is made with the head at all, from the mere whiplash or sheer aggressive force of a hit to the upper body. Yet there is ongoing (and promising) research into new headgear, the Hexlid, that might reduce the risk of concussion.

The codes all continue to invest in scientific research. In women's football, researchers are looking at why women suffer concussion at a higher rate than men – whether the reason is biological (neck strength is a potential determinant), or due to the semi-professional nature of the league, or the greater willingness of women to report symptoms and ask for help.

The AFL has committed $2.5 million per year over the next decade towards a 'substantive longitudinal study of concussion', which includes the use of 'smart' mouthguards, which record and measure linear and rotational head impacts, and should ultimately contribute to a greater understanding of subconcussive hits. This is just one of many studies underway examining the immediate effect of any concussive knock, including everything from blood testing and saliva biomarkers and eye-tracking, in the hope of producing the holy grail for concussion in sport: an objective diagnostic tool, that can be used swiftly on the sidelines. The codes are all sharing awareness programs in the hope that change will also come from amateur and junior leagues. The AFL released the HeadCheck app to help such teams – who don't have paid club doctors – to determine if it might be prudent to remove a player from the field and seek treatment.

Soccer has concussion on its radar, too. The sport's governing body in Australia, Football Australia, has guidelines stating that a player suffering a concussion should be immediately removed from play and can't return until assessed by a medical practitioner – and not on the day of injury. Once cleared to play, they progress through a graduated return-to-play program, from recovery to rehabilitation. It's at least six days before they can play a competitive game. It has also assembled a group of experts to review concussion in the game to ensure best practice, including looking at whether young players should be taught 'heading', which is allowed under international rules. In the United States, children aged ten and under are banned from striking the ball with their heads in both training and matches, while in Britain heading is banned in training for children aged 11 and under, with a graduated approach thereafter. There is no hard and fast rule among clubs in Australia.

Most concussions in soccer are not caused by purposeful headers, though, but by elbow-to-head contact, a head hitting the ground or a goalpost, head-to-head contact or heading 'duels' where two players vie for the ball mid-air but instead clash heads, says sports injury researcher Dr Kerry Peek from the University of Sydney. Using your head to deliberately strike a ball is different to it being hit in a collision, and it remains to be seen whether low-velocity impacts from heading could cause long-term issues with brain health, says Peek, who is on the panel advising Football Australia. There is limited evidence about long-term brain health that distinguishes between episodes of concussion and heading, she says.

Yet, while we don't know if headers cause brain problems down the track, 'we know enough that there could be an issue – so we should do something'. Peek, whose own children play soccer, suggests applying 'the precautionary principle': reducing the number of headers that young players do in training but not removing their ability to learn how to perform a header safely. 'Removing it in training could make it worse because then you'd have players heading a ball in games who don't know how to perform this skill correctly.' This might mean using balloons or much softer balls to improve technique, for example, and instilling the protective value of activating neck muscles.

She also urges a sense of perspective. 'Sport has so many positive benefits,' she says, 'we need to make sure we reduce the risk as much as possible but not take away the opportunity for all the positive benefits [to flow from it].'

When a head knock does occur, in sporting codes around the world a cautious mantra seems to be taking hold, particularly where children are involved: 'If in doubt, sit them out.'

*

Crisis support can be found at Lifeline (13 11 14 and lifeline.org.au), the Suicide Call Back Service (1300 659 467 and suicidecallbackservice. org.au) and beyondblue (1300 22 4636 and beyondblue.org.au)

If you injure your brain repeatedly, it's hardly surprising that it could lead to seizures.

13

WHEN IS THE OPEN PLAN ACTUALLY A TRAP?

More than just a fix for poky rooms and dark houses, the open plan promised light, freedom and minimalist chic in our homes and offices. So how did it end up so often cramping our style?

Elizabeth Farrelly

O ne of architecture's most recognisable images is a dramatic black-and-white shot of two women in big dresses and tiny pumps chatting in a transparent glass box. It's night-time. The glass box is held between two sheer white planes high above the Hollywood Hills. Its ribbed ceiling, uplit, soars out over limitless space. Beyond, gridded with tiny lights and spread to the horizon, is the vast Los Angelean sprawl. That, of course, was before the smog arrived. These days, the same image shows a horizon of brown smudge.

The house is Pierre Koenig's 1959 Stahl House, aka Case Study House #22. In 1945, six months before the end of the Second World War, the young editor of the California-based *Arts & Architecture* magazine, John Entenza, decided to commission radical young architects to design and build new houses, releasing the pent-up energies of wartime in a series of emblematic houses. These houses – by Richard Neutra, Charles and Ray Eames, Eero Saarinen, Craig Ellwood, Koenig and others – changed the world.

Julius Shulman's iconic 1960 photo of Pierre Koenig's Stahl House in Hollywood Hills. Courtesy J. Paul Getty Trust, Getty Research Institute, Los Angeles (2004.R.10)

The Stahl House is a classic. A listed cultural monument, it has featured in countless films, TV shows, music videos and fashion shoots. But in the history of the world, the image – taken by Julius Shulman in 1960 – is bigger. This image, and others like it with their seductive whispers of freedom and power, sold us the open plan. It heralded modern advertising. The two women inside the audacious scarp-edge build seem magically suspended in a world of eternal optimism. Beneath them, concrete floor beams hang in space. In the foreground, a small pot-plant signifies nature tamed, while above, the planar roof flies out in mastery of the Earth itself.

Fast-forward 60 years. The open plan is now a sine qua non of real estate. Every developer apartment, renovated terrace, crappy McMansion and speculative office locates us within yet another space without walls. Too often, with no escape from boss supervision, kid noise, partner grump, cooking smells or the myriad frustrations of COVID lockdowns, we've found ourselves feeling not liberated, but trapped.

So perhaps it's time to wonder whether openness is really the holy grail? Has the open plan – and the minimalist look that goes with it – delivered on its promise? When is an open-plan good, and when is it a trap?

Where did the open plan come from?

The twentieth century was fuelled by versions of freedom: individualism, democracy, creativity, transparency, openness. In retrospect, its failure to fully deliver resonates eerily with Kris Kristofferson's line, immortalised by Janis Joplin in 1971, 'Freedom's just another word for nothin' left to lose.'

The open plan was no mid-century invention. Indeed, when the Stahl House was built in 1959, it was already a half-century old. In the United States, some of the early 'shingle style' architects had begun to experiment with spaces that flowed into one another from the turn of the century. Greene and Greene's legendary

Gamble House in Pasadena (1909) had played with the idea of breaking open the cellular house-plan. With its glazed screens and bays gently held beneath a great maternal roof, it toyed with defining space by use – a dining table or a settee – rather than by walls. People were intrigued and enchanted but such houses were strictly for the wealthy, whose everyday mess could be tucked securely away.

It was the mud and chaos of war that really triggered the craving for a clean, open future for everyone. The first explicit formulation of the open plan as universal principle came from Swiss architect and polemicist Le Corbusier who, in the 1920s, coined the term 'le plan libre' – the free plan. Corbusier's 'Five Points of a New Architecture,' published in 1927, made the 'free ground plan' one of the five essentials of Modern design. The key to it was pilotis (or columns) instead of loadbearing walls.

> *It was the mud and chaos of war*
> *that really triggered the craving for*
> *a clean, open future for everyone.*

That was the domestic world. In office design, a similar reverence for universal openness was reached by a different route. Frederick Taylor was an American engineer obsessed with the idea of efficiency. His theory was known as 'scientific management' or, simply, Taylorism. It merged two new types of nineteenth-century workspace – the commercial office (typing pool or mail-order firm) and the factory production line – into a vast, open-plan work space that meant hundreds of workers could be overseen by one or two supervisors. Not surprisingly, this idea caught on.

Technology was the facilitator. Architecture had always been constrained by the need to put things on top of other things. Walls and columns were necessary to bear the weight; windows had to go between. Steel changed all that. Steel-reinforced concrete made immense spans and cantilevers possible. Floors soared out into space, roofs seemed to float, weightless. Openings could be anywhere, or everywhere. Corridors were obsolete. Enter, the glass box.

The glass box became the dream. Many became architectural icons – from Ludwig Mies van der Rohe's ultra-elegant Farnsworth House in Illinois (1951) to Sean Godsell's very beautiful 1997 Kew House in Melbourne, with no external walls (except glass) and no curtains. As to work, the glass-walled skyscraper – such as the 1955 ICI House at Melbourne's north-east tip by architects Bates Smart McCutcheon or the first AMP building in Sydney, from 1957, by Peddle Thorp – is really just glass boxes piled into tower form.

The ultra-efficient floorplate of the open-plan office enabled the squeezing of space standards and the intensification of supervision. There was, for instance, the phenomenon of the 'typing pool', where 100 – or 200 – (usually) women could be overseen by a single pair of (usually male) eyes. Privacy was virtually non-existent. The clear message was, while you are here, you are 100 per cent company property. In other words, an idea born of the yearning for freedom was rapidly deployed in pursuit of its opposite; oppression. Many workers in open-plan offices today, fighting for a private conversation or phone call, might concur.

The clear message was, while you are here,
you are 100 per cent company property.

Even within a single individual, though, the drives are often contradictory. Modern architecture – indeed, perhaps the whole of modern life – is shaped by the tension between two opposite and insatiable human cravings: for freedom, and for safety. As herd animals, we crave the freedom of the prairie but, at the first scent of danger, scurry back to the safety of the corral. Behind both yearnings – and therefore beneath the open plan – sits fear. And, ultimately, this is the freedom we crave: freedom from the fear of pain and mortality.

Almost all advertising exploits this fear. Modern architecture, in many ways an early product of advertising, is no exception; whetting our freedom appetite without ever fulfilling it. At its best, the open plan offers a sense – perhaps an illusion – of both freedom and safety. The Shulman photo seems to offer limitless view and openness in all directions with no loss of privacy or control. There is allure here. But it's the allure of the unachievable.

Mies and Godsell are both very fine architects. Yet Dr Edith Farnsworth sold what may be the world's loveliest house after complaining of never being able to hide the bins, keep mosquitoes out or conceal dirty dishes.

Has the open plan delivered on its promise?

The history of Modern architecture is twined in ironies. One is that the open plan, which began as a push for universal equality and freedom, has become yet another device to entrap the poor and enrich the wealthy. What began as a critique of orthodoxy became its tool.

In the 1920s and the 1950s, the immediate post-war periods, Modernism strove to break free of historical constraints, including those of inherited wealth and class. So it lionised typologies such as schools, health centres, public and affordable housing that would benefit everyone and adopted aesthetic values – simplicity, plainness, openness – that seemed similarly aligned. Even residential towers carried this idealism. In theory, gathering ground-hugging slums into a single tower left the ground that had been vacated open for park-like gardens. There, it was said, children could play while mothers supervised from the tenth-storey kitchen window.

Not surprisingly, with such a sales blurb, the open plan quickly gained popularity.

This socialist drive was everywhere. In Auckland, in 1949 and 1950, a bunch of architecture students calling themselves the Group Architects built two shed-like houses as habitable manifestos. Design was a socialist platform. Known simply as First House and Second House, these dwellings, with almost no internal walls, were intended to make usable, delightful, climate-appropriate shelter available for all. Architecture, polemicised the Group, 'can only arise out of the daily life of Everyman, and without Everyman there can be no architecture'.

In a world bursting with optimism, the dream was irresistible. Philip Johnson's own all-glass house set in an oak park in Connecticut (1949) was hugely influential, and Mies's Farnsworth House even more so, as were the Case Study houses. In Australia, the glass-house dreaming manifested itself in works by Harry Seidler, Robin Boyd, Bill Lucas and others.

> *In a world bursting with optimism,*
> *the dream was irresistible.*

But there were problems. A house without walls is one thing on the Hollywood scarp or a 10-hectare estate, and quite another on a tiny suburban block. An open plan is one thing if you are across its socialist polemic; another if you just want to be able to put the kids' toys away.

The glass box was an Edenic dream. Adam and Eve could be naked in paradise because they were clothed in solitude and innocence. Replicate those Edens, set them cheek by jowl, send

Adams and Eves, suited and brief-cased, off to work each day to feed their growing tribes, and the picture is altogether different. In that situation, the need for privacy demanded that houses be separated by space and planting, helping to entrench the low-density, car-dependent sprawl that typified the twentieth century across the world, and all of its environmental detriments.

Now, the open plan is ubiquitous. Virtually every habitable building from the past 20 years boasts a farmhouse kitchen, open-plan living or open workspace. And true, there are upsides to this. You can't enter a Victorian terrace house or Federation bungalow without wanting to blow the back off it, replacing the dank huddle of laundries and outhouses with a big, bright, airy space that opens the shared life of the household to a garden or court.

At the same time, cracks are starting to show. We're seeing, for example, the advent of the 'mess kitchen', 'frying kitchen' or 'butler's pantry' – where the real work, and the real mess, is tucked away while the ultra-schmick, detail-free island kitchen is there for everyone to see.

In the hands of developers, further, openness has become too often an excuse for meanness, so that 'living–dining' is barely two-thirds the size it would be as two rooms, while the study can be a tiny under-stair desk and the bedroom merely an alcove off 'living'. The pandemic has highlighted the way such open-plan living can become an acoustic and emotional nightmare, with noisy toddlers, television, online meetings and family hubbub competing for mastery of a single space.

Of course, there's also the energy question. Big open spaces are energy guzzlers when it comes to heating and cooling, and significant contributors to climate catastrophe. Wise architects, therefore, make thermal comfort a priority – houses by Richard Leplastrier and Peter Stutchbury spring to mind – and will often provide shutters, sliding doors or heavy curtains as seasonal space dividers.

Is our attachment to objects unhealthy?

An unspoken subtext of the open plan is minimalism. Essential to the look is a clean, uncluttered floor plane that flows seamlessly into a garden, infinity pool or view. It's all about the impression of limitless expanse. And, as space diminishes, the push to declutter has acquired a whole new urgency.

Enter Marie Kondo, the Japanese tidiness queen whose folding advisories have made her a global celebrity. Kondo's advice about keeping only those things in your life that make you smile is endearing, possibly even wise. And it's true that, in an open plan, storage is the key to survival.

But decluttering can go too far. Tidiness is the Botox of domestic life; a desire to iron out the wrinkles, to render it perfect. (I once knew a woman who told me, in all seriousness, that she had spent 50 years trying not to laugh or smile too much because it would give her wrinkles. She was old and I was young, but I felt the tragedy of it even then.)

Tidiness is the Botox of domestic life; a desire to iron out the wrinkles, to render it perfect.

The pursuit of perfect order has the same effect on our domestic lives. I'm not immune to the charms of a big, sunny space, but I also want options. To live without mess or randomness is to live without creativity, spontaneity or genuine emotion. An abiding emotional attachment to objects for their significance or beauty is surely healthy; an exercise in embodied cognition. And to recognise their spirit-content is also to see the beauty in the fact that such objects have a life of their own. It's not all about control. Investing objects with emotional significance should also limit our easy-come-easy-go material consumption.

Importantly, it should also enhance our respect for the material environment we call nature more than a few random status-object coffee-table books artfully placed.

Naturally, this is not everyone's view – but that's the point. Rather than losing all our objects, we should strive to acquire (or keep) only things we can actually love; things that have meaning or significance, or that give us genuine joy. We should never buy a house or apartment for the glamorous photo. Rather we should recognise archi-porn for what it is – a form of propaganda.

When is an open plan good, and when does it become a trap?

In the end, it's about fit. The dwelling, be it house, apartment or shoebox, is a garment. It needs to fit you and your co-dwellers for size but also in terms of shape, style and messaging. And very little of that figures, even for an instant, in the housing-delivery system that we have.

What has emerged from the recent flood of spec-built apartments and office blocks that over the past decade have deluged Australian cities is a traduced, soulless and depressingly uniform version of the open plan. Just as the open office was embraced by early-twentieth-century capitalists as a means to more efficient exploitation, so this century's developers have eagerly recast the domestic open plan as a way to cut costs and maximise profits; an approach encouraged by deregulatory neoliberal planning. It was Mies van der Rohe who coined minimalism's famous aphorism 'less is more'. And sometimes it's true, at least up to a point. Just as often, though, especially in the greedy hands of developers, less is simply less.

The cost-savings arise because a kitchen or study that, as a single room, would seem impossibly mean looks much more plausible enveloped within a larger space. A workspace that would seem skimpy as a cell feels okay when its boundaries are removed or blurred. Further, the fewer walls and doors you have to build, the lower the outlay. But the result, ironically, can be oppressive.

Traditional, room-based architecture can be cold and dark. It can be confining, especially when badly designed. But it also offers complexity and intricacy. It offers options, nooks and crannies. Ironing out these wrinkles and invaginations has been one of modernism's enduring projects. This occurred at the large scale of the city where, from mid-century on, crooked lanes and narrow alleys were deliberately erased for open malls and plazas. It occurred also in ecclesiastical architecture, where the cloisters and side-chapels of Gothic architecture were replaced by the all-inclusive, big-space open plan, where there is no longer any place to withdraw, to grieve or simply to worship in solitude.

And so it is in domestic architecture, especially in the extreme version of the open plan. As Godsell says of his Kew House, 'Because this building forces one to confront oneself, then if you don't really feel good about yourself then you probably don't like the building.' I'm a Godsell fan, but this is architecture's hubris – that people must adapt to it, rather than vice versa.

In truth, a good open plan is one that fits you – not just some glamorous, dinner-party idea of who you are, but the truth of you and your tribe. To this end, we should be – and should be encouraged to be – far more involved in the creation and delivery of our own dwellings via co-operative or crowd-funded building projects. The idea that a house or flat is a commodity, manufactured by a profiteer to some lowest-possible standard, has undermined some of our deepest values. A dwelling, open plan or not, should engage you in a genuine give-and-take relationship. It should be an object not of status or show, but of love.

14

HOW DO YOU WIN AN OSCAR?

They're the shiniest awards
in the film business. What goes into
deciding the Academy Awards?

Garry Maddox

The Academy Awards are the ultimate prize in the movies, an annual celebration of creativity, celebrity, glamour and style that has been showcasing the best in cinema for more than nine decades. The often overwrought acceptance speeches show that winning an Oscar can be an overwhelming experience for even famous Hollywood directors and actors, let alone unknown visual effects technicians, make-up artists and short-filmmakers.

Oscars are, of course, hard to win. Some of the world's greatest auteurs – Orson Welles, Alfred Hitchcock, Stanley Kubrick, Howard Hawks, Ingmar Bergman, Akira Kurosawa – have never won best director. And your chances have been markedly less over the years if you were a woman. Only two have won best director – Kathryn Bigelow (*The Hurt Locker*, 2009) and Chloe Zhao (*Nomadland*, 2021).

> *If you're an actor, find a story about someone famous and, ideally, beloved.*

To win an Oscar, you could rely on decades of dedication to your craft, hard work, smart career decisions, industry connections and plain good luck . . . but there are ways to boost your chances.

You could direct a movie for Pixar, for example, given the studio has won best animated feature a remarkable 11 times in the past 18 years.

You could become a key member of a sound team on a big war, action, sci-fi or music movie given those genres have tended to triumph.

If you're an actor, find a story about the struggles of someone famous and, ideally, beloved. Think Renee Zellweger as Judy Garland, Rami Malek as Freddie Mercury, Gary Oldman as Winston Churchill . . . And, historically, it has helped to be white.

Winning best picture at the Oscars is like winning the lottery. It can be highly rewarding but you have to beat the odds against thousands of movies released around the world every year.

But what goes on behind the scenes?

How do movies qualify for best picture?
A movie has to have at least a seven-day run in Los Angeles county in the previous calendar year – at least three times a day including once at night – to qualify. (This rule was suspended in 2020 due to COVID.) The voters are members of the Academy of Motion Picture Arts and Sciences, whose branches cover actors, directors, writers, producers, cinematographers, editors and other industry professionals. They can all vote, firstly, for the best picture nominees then the winner.

Six years ago, there were 6261 members but the outcry over the lack of any people of colour nominated in the four acting categories sparked a movement for change with the social media hashtag #OscarsSoWhite. When it happened again the next year – with a survey finding the membership was 94 per cent white and 77 per cent male – the Academy pledged to double the number of women and ethnically under-represented members by 2020.

*Winning best picture at the Oscars
is like winning the lottery.*

As a result, membership has grown to more than 9900 and the Oscars have seemed more diverse, with a stronger racial mix among the acting, directing, best picture and screenplay nominees. And after only five women were nominated for best director in 92 years, two were up for the first time in 2021.

In another step towards change, the Academy introduced inclusion standards for a movie to be eligible for a best picture nomination. They say it must meet two of four criteria, having cast and crew from minority groups and featuring storylines or themes centred on under-represented groups. The number of nominees has been increased from five to as many as ten. To be nominated, a movie has to get at least 5 per cent of first-place votes. As with Australian elections, the winner is determined by preferential voting.

A distributor or streaming service will often schedule a premiere for at least one of three major festivals in September – Venice in Italy, Toronto in Canada and Telluride in Colorado. They hope positive reviews and industry talk will result in their movies being selected for other festivals, awards and the many lists of Oscar contenders that bob up in the media. Well-funded marketing teams will get to work, but much of the campaigning goes on behind the scenes.

How much does campaigning matter?

While the Oscars have always been more serious and upstanding than the Hollywood Foreign Press Association's Golden Globes – Ricky Gervais regularly joked about its ethical standards while hosting – aggressive lobbying has sometimes crossed the line and has led to the tightening of campaign rules.

The now disgraced producer-distributor Harvey Weinstein ran what he called guerrilla campaigns that *Forbes* magazine calculated contributed to 341 nominations and 81 wins from 1990 to 2016, including best picture for *The English Patient*, *Shakespeare In Love*, *Chicago*, *The King's Speech* and *The Artist*.

Weinstein's campaign tactics included paying veteran publicists who were Academy members to schmooze other members, spending US$5 million to campaign for *Shakespeare in Love* at a time when studios were typically spending US$2 million, running smear campaigns against rivals, even getting a publicist to write an opinion piece praising Martin Scorsese for *Gangs of New York* in the name of *Sound of Music* director Robert Wise (with Wise's cooperation).

Is campaigning all above board now? There are certainly strict rules but an Australian Academy member in the United States has this to say about the restriction on wining and dining voters: 'Pre-COVID, Academy members could count on at least two months – four days a week – of lavish dinners at exclusive restaurants with the cast and director following a cinema screening of a contending movie. With the pandemic, the dinner comes to your house, with Netflix offering US$100 vouchers for UberEats plus care packages such as very expensive whisky along with two crystal glasses and special chocolates.'

Streaming service Netflix, which has had more Oscar nominations than any other studio or distributor for the past two years, has become the new campaigning heavy-hitter. Even though an executive denied a report that it spent more than US$100 million campaigning in 2020, it is widely believed Netflix at least matches other studios' US$5 million to US$20 million per movie and has more of them in contention.

As well as extensive advertising – TV, print, online and billboards – awards campaigns traditionally include screenings, parties and travel for contenders to events and festivals, both in the United States and internationally.

In more conventional marketing terms, they will craft narratives to appeal to voters. When Leonardo DiCaprio did interviews for the western *The Revenant*, for example, he talked about shooting in freezing temperatures, spending up to five hours a day in the make-up chair to get prosthetics that looked like bloody wounds, repeatedly diving into an icy river, eating raw bison liver and sleeping in animal carcasses during filming. After losing with four previous nominations, DiCaprio had a compelling case to win ... and he did.

A powerful narrative can swing votes for best picture as well. In 2020, there was a strong case that a more inclusive Academy should make history by crowning South Korea's *Parasite*, a black comedy about social inequality from revered director Bong Joon-ho, as the first foreign-language winner of best picture. It won.

> *A powerful narrative can swing votes*
> *for best picture as well.*

Other times the narrative works against a favoured nominee. In 2019, Netflix's *Roma*, a drama about a maid holding together a struggling family in Mexico City, was widely recognised as masterful filmmaking, but an argument emerged during campaigning that the Academy should not reward a streaming service that was trying to destroy the cinema business. The result: a surprise win for the racially charged road movie *Green Book*.

Two years earlier, there was an argument that the favoured *La La Land* – a popular musical about a pianist and an actress falling in love – did not deserve to win because it was about 'a white guy who wants to save jazz' and that it failed on racial and gender politics. There was also a Twitter storm about it lacking gay characters. The result: a shock win for *Moonlight*, an edgy drama about a young gay African-American man.

What type of movies win best film?

In 1929, the first award for what was then called outstanding picture went to *Wings*, a silent epic about two rival fighter pilots in the First World War. The ceremony, at a private dinner at the Hollywood Roosevelt hotel, lasted just 15 minutes.

A musical, *The Broadway Melody,* won the next year, when the event at the Ambassador Hotel was broadcast on radio. The next year it was back to war movies again, for *All Quiet on the Western Front* (and this time the ceremony was filmed). This was around the time the award became known as an Oscar after, as the story goes, an academy secretary said it reminded her of her Uncle Oscar.

Certain types of movies have been perennially out of favour when it comes to best picture. A sci-fi, superhero, animation or children's movie has never won and only rarely has a comedy, comic-drama or horror-thriller triumphed.

Winners have included many movies that went on to be regarded as classics, including *Gone with the Wind, Casablanca, On the Waterfront, Lawrence of Arabia, The Sound of Music, Midnight Cowboy, The Godfather* and *The Godfather Part II*.

But other classics have missed out. *Citizen Kane* lost to *How Green Was My Valley; Taxi Driver* was pipped by *Rocky; Star Wars* was beaten by *Annie Hall*.

But unlikely movies still come out on top.

It would have been unimaginable to Bong Joon-ho when he was shooting *Parasite* – with dialogue in Korean and a cast unknown in the United States – that it would be the first foreign-language winner of best picture in 2020. Almost as unlikely would have been the 2012 win by a black-and-white silent movie from a little-known French director in *The Artist*.

Australian producer Emile Sherman, who won in 2011 with *The King's Speech*, says that every Hollywood studio originally passed on the drama about how a speech therapist helped the stuttering future King George VI, dismissing it as 'two men talking in a room'. After being nominated again for *Lion*, a drama about an adopted Indian-Australian man searching for his birth mother, he has realised how much winning best picture is a lottery.

'It's an incredibly chaotic process with a huge amount of luck involved and you just have to land at the right place at the right time,' Sherman says.

If a movie examines a great life, historic event or burning social issue, and if it deals with such powerful themes as life, death, love, duty, truth and sacrifice, its best picture chances are better than if it has, say, Godzilla fighting Kong.

If it's a dramatic story about people at the top of their field, that will help, too. They could be entertainers (*Shakespeare In Love*, *Amadeus*), royalty (*The King's Speech*, *The Last Emperor*), soldiers (*The Hurt Locker*, *Gladiator*, *Braveheart*), athletes (*Million Dollar Baby*, *Chariots of Fire*, *Rocky*), brilliant and unorthodox thinkers (*A Beautiful Mind*, *Rain Man*, *Forrest Gump*) or even journalists (*Spotlight*). But lately it helps to focus on humble people struggling to make a life (*Moonlight*, *The Shape of Water*, *Parasite*, *Nomadland*).

On a deeper mythic level, many best picture winners have been about an individual overcoming apparently impossible odds, with unexpected assistance, to achieve something important (*The King's Speech*, *Slumdog Millionaire*, *The Lord of the Rings*, *The Return of the King*). If the hero dies (*Titanic*, *Braveheart*, *Gladiator*) or is damaged (*The Hurt Locker*, *American Beauty*), something positive must come from the sacrifice.

The best picture award goes to the producers of a movie – who oversee the production – rather than the director, who is the main creative force. Given there are often many producers, executive producers and co-producers, only the key ones are included.

Are other awards a good predictor?

The guilds that represent Hollywood producers and distributors are statistically significant – though far from infallible – in predicting what will win best picture. For the Producers Guild of America Awards, it has happened about two-thirds of the time. And for the Directors Guild of America Awards, the winning director's movie has also won best picture about two-thirds of the time.

One factor that is irrelevant is having the most Oscar

nominations. It has become a statistical rarity for the most-nominated movie in all categories to win best picture.

And sometimes the Oscar race is never over, even when it's over. As *Moonlight* showed during the 'wrong envelope' fiasco in 2017, a movie can still win best picture more than two minutes after the award has been given to another nominee. As for other predictors, the BAFTAs and Golden Globes only rarely foreshadow best picture success at the Oscars.

What difference does it make anyway?

A best picture win can open up opportunities for producers and directors, boosting their status in the industry and making others keener to work with them. The adjective 'Oscar-winning' can open doors even if such recent winners as producers Mollye Asher (*Nomadland*), Kwak Sin-ae (*Parasite*) and Jim Burke (*Green Book*) are not exactly household names. But much depends on what else producers want to do for the rest of their careers.

Studios and distributors spend so much to win best picture because they hope to get it back from a boost to box office takings (if the movie is still showing) and later from sales for online, television and other releases.

Streaming services know the cachet of just a nomination can attract new subscribers and keep existing ones. Netflix, for example, reported that households watching *Mank*, the black-and-white drama about the writing of *Citizen Kane*, jumped more than 700 per cent after it became the most-nominated movie at the 2021 Oscars.

But for a filmmaker, there are other less tangible ways of measuring success. If a movie has touched lives, drawn attention to an important issue, revealed a story the world needs to see, inspired someone, kickstarted a brilliant career, brought joy or even just been great entertainment, who cares about awards? The work is the reward.

15

HOW DO YOU FIND A DALAI LAMA?

He was two when monks found him in a village in Tibet. But how did they know he was the Dalai Lama? And why will superpowers one day wrangle over his successor?

Matt Wade

He's been at the helm of Tibetan Buddhism for eight decades. But the Dalai Lama reckons he'll be around for at least two more. The spritely 86-year-old often cites an eighteenth-century prophecy that he, the fourteenth Dalai Lama, will live to the age of 113. Other spiritual signs back up that divination. 'I've had dreams about living long,' the Dalai Lama told followers in 2019. 'In one dream I was climbing steps, 13 steps, which I interpreted to relate to the prediction that I could live to the age of 113.'

Despite these assurances, there's speculation about what will happen when the time comes to find the Dalai Lama's successor. The answer is not so simple. A process steeped in mysticism, which once played out in relative obscurity on the Tibetan plateau, will next time attract global attention. The Dalai Lama's death is likely to herald a struggle between China and Tibetans over who controls Tibetan Buddhism. And, while the Dalai Lama describes himself as a 'simple Buddhist monk', even the anticipation of his reincarnation can get superpowers squabbling.

> *A process steeped in mysticism, which once played out in relative obscurity on the Tibetan plateau, will next time attract global attention.*

How was the Dalai Lama chosen for his job? Why do the Chinese call him a 'wolf in monk's robes'? And what is at stake in choosing his successor?

Who is the Dalai Lama?

The monk known for his trademark saffron and claret robes is one of the world's most recognisable religious figures. The Dalai Lama, who is referred to by followers as His Holiness, won the 1989 Nobel Peace Prize for 'advocating peaceful solutions based upon tolerance and mutual respect'. Since China's occupation of Tibet 70 years ago, he has come to symbolise the struggle of his people known globally by the iconic slogan Free Tibet.

Tibetan Buddhists believe all beings come to this present life from a previous one and that they will be reborn again after death. Several hundred special reincarnational lineages have been identified in Tibet, the most respected being that of the Dalai Lama.

The current Dalai Lama is the fourteenth in a line of monks going back five centuries who are believed to be incarnations of the revered Avalokiteśvara or Chenrezig – a Bodhisattva of Compassion described as the patron saint of Tibet. Bodhisattvas are beings 'inspired by the wish to attain complete enlightenment, who have vowed to be reborn in the world to help all living beings,' says the Dalai Lama's official description.

Dalai Lamas have played a dominant role in Tibetan politics in addition to their spiritual leadership. Dalai Lamas, or their regents, governed much of the Tibetan plateau with varying degrees of autonomy from the seventeenth century until China occupied Tibet in 1950.

After he was 'discovered' (more on that later), the current Dalai Lama was enthroned in 1940 and undertook monastic training in Tibet, achieving the highest doctorate in Buddhist philosophy. But he fled his homeland in 1959 after a Tibetan rebellion against Chinese rule was forcefully repressed. Tibet has undergone sweeping changes in his absence. China has invested heavily in the development of the region but human rights groups claim Chinese assimilation policies are destroying Tibet's unique cultural identity.

For more than 60 years the Dalai Lama has lived in a compound with a garden where he walks each morning if it is not raining, near a Tibetan Buddhist temple in Dharamshala, a picturesque Himalayan hill station in North India. The town hosts the Central

Tibetan Administration, also known as the Tibetan government in exile, which represents hundreds of thousands of Tibetans who have fled their homeland since it was annexed by China.

Since the 1980s, the Dalai Lama has advocated a 'middle way approach' to finding a negotiated resolution with China over the Tibet issue. The policy seeks to gain 'genuine autonomy' for Tibetans while the region remains under the Chinese government. But Beijing rejects this – it has labelled the Dalai Lama a 'wolf in monk's robes' and considers him a dangerous separatist seeking to split Tibet from China.

I first encountered the Dalai Lama at a difficult moment for him and the Tibetan people. It was March 2008 and anti-China protests inside Tibet had turned violent, attracting global media attention. I was among a small group of journalists from around the world who assembled at the Dalai Lama's residence to hear his take on the bloodshed. The press conference was sometimes tense as reporters questioned why he was not being more critical of the repressive tactics being used by Chinese authorities inside Tibet.

At one point, the Dalai Lama leant forward in his chair, fixed me with an intense gaze and described the predicament facing all Tibetans campaigning for greater autonomy for their homeland. 'In the last few days I have this sort of feeling of a young deer in the tiger's grip,' he said, using his hands to mimic claws. 'Can the deer really fight with the tiger? It can only express the truth. Our only weapon, our only strength, is justice and truth.'

The Dalai Lama also defended his middle way approach to China. 'We must be practical,' he said.

The conclusion of that press conference was different to any I have experienced. The Dalai Lama stayed on to greet his inquisitors one at a time. With charm and grace, he shook hands, joked and chatted. Reporters now jostled to meet the famous monk they had been grilling minutes earlier. When my turn came, the Dalai Lama held my hand for some time and remarked on how far I had come to attend. I left with a deeper appreciation of why this Dalai Lama is so widely admired.

How is a Dalai Lama selected?

When a Dalai Lama dies – or even before their death – a successor is found rather than chosen. Traditionally, senior Tibetan monks conduct an elaborate quest to find a child who is the Dalai Lama's next incarnation. The search involves consulting oracles, interpreting visions and reading spiritual signs.

> *When a Dalai Lama dies – or even before he dies – a successor is found rather than chosen.*

The senior monks may find clues from the deceased body of the Dalai Lama such as the direction it faces or its posture. If the body is cremated, the direction of the smoke is monitored as a potential indicator for the direction of rebirth.

Dreams are an important guide. Those involved in the search for the Dalai Lama's reincarnation often meditate at Lhamo La-Tso, an oracle lake in central Tibet, and wait for a vision or insight into the whereabouts of the Dalai Lama's rebirth.

Once these visions and signs have been followed up and a potential child found, there is a series of tests to verify the rebirth. The child is presented with artefacts, some of which belonged to the previous Dalai Lama. If the child identifies which objects belonged to the Dalai Lama, it is taken as a sign.

The current Dalai Lama was found at the age of two after a senior monk saw his village and house in a vision at the oracle lake. The boy – then named Lhamo Thondup – was able to pinpoint artefacts that had belonged to the previous Dalai Lama, including a drum used for rituals and a walking stick. 'It's mine, it's mine,' he is reported to have said.

The boy was then educated by monks and prepared for his lifelong monastic role. In 1940, when he was five, he was officially installed as the spiritual leader of Tibet at the Potala Palace in the Tibetan capital, Lhasa, and renamed Tenzin Gyatso.

The current Dalai Lama, aged four in 1939
Courtesy: A.T. Steele Papers, Special Collections,
Arizona State University Libraries

Will finding the next Dalai Lama be different?

Identifying the next Dalai Lama will be especially tricky because politics is destined to complicate the search.

The Dalai Lama has hinted – often in a light-hearted way – that he could reappear anywhere, in any number of human incarnations, possibly as a woman. He's even speculated that the

Dalai Lama may no longer be needed (although he says that's up to the Tibetan people).

In a 2011 speech about reincarnation, the Dalai Lama said that at 'about 90' years of age he will consult religious leaders, and the Tibetan public, to make a final decision about the selection of his successor. The primary responsibility for identifying his reincarnation will rest with officers of the Dalai Lama's Gaden Phodrang Trust, which includes his most senior monastic disciples. He also vowed to leave 'clear written instructions' on the process.

But the Chinese government already seems to have made its own plans. It claims the legal right to recognise Tibetan reincarnations. When the time comes to select a new Dalai Lama, China is expected to handpick its own candidate in a bid to ensure Tibet's top Buddhist leader is friendly to Beijing. 'Reincarnation of living Buddhas including the Dalai Lama must comply with Chinese laws and regulations and follow religious rituals and historical conventions,' Chinese Foreign Ministry spokesperson Geng Shuang said in 2019.

> *When the time comes to select a new Dalai Lama,*
> *China is expected to handpick its own candidate.*

A precedent was set during the last transition to a new Panchen Lama, the second-most important figure in Tibetan religion, culture and politics.

In May 1995, the Dalai Lama himself recognised a six-year-old living in Tibet named Gedhun Choekyi Nyima as the new incarnation of the Panchen Lama. But the boy disappeared soon after and human rights groups say the Chinese government has kept him and his family in a secret location ever since.

The Chinese government announced the son of two Communist Party cadres, named Gyaincain Norbu, was the next reincarnation of the Panchen Lama.

Deakin University professor John Powers, who is a scholar of Tibetan Buddhism, says most Tibetans do not consider China's choice to be legitimate.

'They think it's ridiculous,' he says.

Even so, the scene is set for a complex struggle following the death of Tibet's exiled spiritual leader.

The Dalai Lama himself has predicted his next incarnation will not be found in Chinese-controlled Tibet anyway. 'My life is outside Tibet, therefore my reincarnation will logically be found outside,' he told *Time* magazine in 2004.

Chope Paljor Tsering, a former minister in Tibet's government in exile who now lives in Canberra, says it makes spiritual sense for the Dalai Lama to be reincarnated beyond the control of Chinese authorities.

'The very purpose of the next reincarnation is to continue the legacy of the previous person,' he says. 'Therefore, if the next Dalai Lama is to be found while the current one is in exile still struggling for the freedom of the Tibetan people, and the preservation of Tibetan culture and religion, then obviously he will be reincarnated in a free country. That includes anywhere in the world where there is a sizeable Tibetan Buddhist community . . . you might even have an Australian Dalai Lama.'

However, the most likely non-Tibetan location for the Dalai Lama's next incarnation is India, where more than 100,000 Tibetans live, most of them refugees.

What's more, a Dalai Lama has previously been found within the borders of modern India. The sixth Dalai Lama was born in 1680 at Tawang in the eastern Himalayas, not far from India's border with Tibet. The town's imposing hilltop monastery is an important centre for Tibetan Buddhism. This precedent means Tawang would be a highly credible location for the next Dalai Lama to be found.

Another alternative canvassed by the Dalai Lama's supporters is that he identifies a reincarnation while he is still alive, possibly a well-credentialled teenage monk.

Chope Paljor Tsering claims this option would be popular with Tibetans. 'It will be very easy for Tibetans to accept such a concept,' he says. 'There have been precedents for this – not with the Dalai Lama but with other very high-ranking religious leaders.'

But Powers says that recognising a reincarnation 'while the incumbent is still alive' has been rare in the Tibetan Buddhist tradition.

While the method for identifying the Dalai Lama's next incarnation remains uncertain, the transition promises to be contentious.

Dr Ruth Gamble, a La Trobe University academic who has studied the reincarnation traditions of Tibetan Buddhism, says the world is likely to end up with two new Dalai Lamas – one endorsed by the Chinese government, the other chosen by Tibetan monks in exile.

'What you've basically got is two institutions claiming legitimacy to the process of the Dalai Lama's reincarnation,' she says.

How will Tibetans respond?
It is difficult to know how the millions of Tibetans who live in China will react to the controversies that seem inevitable in the transition to the next Dalai Lama.

Critics of China's policies in Tibet say censorship and propaganda mean a growing number of Tibetans are growing up with limited knowledge of their own culture and customs.

But Gamble is worried the Dalai Lama's death could trigger unrest inside Tibet reminiscent of widespread protests in 2008, which were forcefully quelled by Chinese security forces.

Tibetans are growing up with limited knowledge of their own culture and customs.

'I'm really scared about what is going to happen on the ground in Tibet when the Dalai Lama dies,' she says. 'I don't know what people will do.'

Powers believes any candidate for Dalai Lama endorsed by Beijing will not gain widespread acceptance. 'The Chinese government is going to name a successor but no Tibetan Buddhist is going to accept this as valid,' he says.

Map by Matthew Absalom-Wong

The Dalai Lama himself warned followers in 2011 not to recognise any candidate for his reincarnation 'chosen for political ends by anyone, including those in the People's Republic of China'.

He has also acknowledged the likelihood that two successors are named.

'In future, in case you see two Dalai Lamas come, one from [India], in a free country, and one chosen by the Chinese, then nobody will trust – nobody will respect (the one chosen by China),' he told Reuters in 2019. 'So that's an additional problem for the Chinese. It's possible, it can happen.'

Chope Paljor Tsering says the Tibetan community is well prepared for the transition to a new Dalai Lama, whatever happens.

'I think people are very conscious of the fact that there is a strong, determined Chinese authority that will do whatever it takes for them to get control over the next Dalai Lama,' he says. 'But the Tibetan Buddhist tradition is strong enough to withstand such action by an atheist authority so, in the long run, I think it is going to be okay.'

Why would choosing a new Dalai Lama stoke tensions?

In December 2020, the US Congress passed landmark legislation reaffirming the rights of the Tibetan Buddhists to choose the Dalai Lama's next incarnation without interference from China. The Tibet Policy and Support Act, signed into law by then-president Donald Trump, empowers the US government to impose sanctions on Chinese government officials who try to interfere in the process of selecting the next Dalai Lama.

In a speech supporting the law, US Speaker of the House of Representatives Nancy Pelosi said, 'If we don't speak out for human rights in China because of commercial interests, then we lose all moral authority to speak out for human rights in any other place in the world.'

The new law infuriated Beijing. The Chinese Foreign Ministry immediately accused the United States of 'meddling in China's internal affairs' and warned the move could further harm 'cooperation and bilateral relations'.

The identification of a new Dalai Lama also has the potential to imperil relations between Asia's giant, nuclear-armed neighbours China and India. If Tibetan monks in exile find the Dalai Lama's reincarnation within India, possibly Tawang, it would provoke an angry reaction from the Chinese government and feed into existing tensions between Beijing and Delhi.

'That would drive the Chinese absolutely berserk,' says Powers.

A popular reincarnation of the Dalai Lama living outside China could fuel another generation of agitation for Tibetan autonomy.

India's willingness to host the Dalai Lama since he fled Tibet has vexed ties between the world's two most populous countries for more than six decades. 'The Chinese government has always protested to the Indian government for allowing a Tibetan government in exile to operate there,' says Powers.

Relations between Beijing and Delhi deteriorated sharply in 2020 following deadly hand-to-hand clashes between Indian and Chinese troops at the disputed border between the two nations in the western Himalayas.

The transition to the next Dalai Lama looms as a flashpoint for India–China relations, says Ian Hall, professor of international relations at Griffith University. 'This is something that causes acute anxiety in Delhi. His death is inevitable but when it happens it will cause problems.'

It's difficult to know exactly how the transition to the next Dalai Lama will play out. But when the time comes, the world will be watching.

16

WHAT DO SHARKS WANT (AND WHY DO THEY BITE)?

A spike in shark attacks in any year always gives rise to the (nervous) question, why? What do we know about the secret lives of these ancient creatures?

Sherryn Groch

Suppose your worst-case scenario goes like this: by the time you turn, he's already seen you, and he's closing fast.

You've seen him before, too, this great white. Not in the surf, but when you close your eyes, when you glimpse a shadow in the water, this is the shark you imagine: 2000 kilograms of muscle and teeth with the black, dead stare of a killer.

Now he's really here, and perhaps you're close enough to see that his eyes aren't black at all, they're navy; dark in the deep but in the sun they shine inky blue. His teeth catch the light, too. You might see them first, jagged like steak knives.

Of course, more likely than not you see nothing but a flash of grey before the bite. It's gentle, by shark standards.

Your leg jerks. You kick out, throw a punch. He lets go.

Now, above the surface, tragedy is unfolding. Lifeguards and panicked swimmers rush to help. But the shark didn't stick around. He's already gone, back into deeper water, with a strange taste in his mouth. So, what did he want?

Great whites travel thousands of kilometres along Australia's coastline and beyond every year. Some are tracked by researchers; they have names, or at least numbers, and their own radio frequencies, thanks to tags implanted under their skin and dorsal fins. This shark doesn't have a tag yet. We'll just call him Shark 26. (That's how many years he's been alive.) For a great white, he's barely grown up, one of more than 8000 whites estimated to be left in Australian waters.

Not all of the 500-odd shark species out there look like Shark 26. Some don't even look like sharks – our native wobbegong more closely resembles a carpet while the Greenland shark grows large and lives for hundreds of years in the icy waters at the top of the world.

In the past few decades, tracking programs have started to piece together a new picture of the ocean's most fearsome predator. Scientists have learnt that sharks, once considered notoriously solitary, are surprisingly social; that some will return to the same stretch of coast alongside the same sharks year after year; that they can learn from each other, remember, and even count; that they might lean in to human touch, lay their head in the lap of a diver or play in the bubbles of their tank.

> *Just over 80 people a year have run-ins with sharks in so-called 'unprovoked attacks'.*

Researchers also have more than 100 years of shark attack records to help calculate the odds of your worst-case scenario ever becoming a reality. So, with 2020 the deadliest shark year since 1929 (when panicked Sydneysiders advocated dynamiting sharks at Bondi), what do we really know about why these big fish bite?

What are the odds?
You're more likely to be killed by bees or cows than sharks, says Gavin Naylor, an evolutionary biologist who curates the International Shark Attack File. On average worldwide, just over 80 people a year have run-ins with sharks in so-called 'unprovoked attacks', when the shark seems to come from nowhere, rather than being lured by fishing or bait. About six people usually die.

Consider that number against the many millions of people taking to the water. Or against the tens of millions of sharks killed each year for their meat and fins as industrialised fishing pushes a third of all shark species towards extinction. Even the great white is now endangered.

Still, bites are no less terrifying or tragic for being rare. In 2020, eight people died in Australian waters alone. That's a very bad year, Naylor says, though not necessarily a trend. There were no bite deaths at all in Australia in 2019.

The great white (also known more humbly as the white shark) has two genetically distinct populations in Australia: western whites range from Victoria to Western Australia while an eastern group moves from Tasmania to Queensland, shifting north in winter. That's similar to the annual movement of whales along the east coast, but it's not a straightforward chase. They can live to the age of 70.

Illustration by Jamie Brown

The 2020 tally 'really hurts', says surfer Dave Pearson, who has been running the world's only shark attack survivor support group, Bite Club, since his own run-in with a bull shark in northern New South Wales almost ten years ago. 'Whenever you chat to a surfer these days, inevitably the conversation turns to sharks.'

In the statistics, you can see the curve of shark bites climbing globally since proper record-keeping began in 1958 (and a particular rise in Australia since the turn of the century). And herein lies the mystery for scientists: if shark numbers are going down, why are attacks going up?

Naylor offers two practical reasons: there are more people in the water than ever before, and there are more people reporting encounters than ever before. 'Fortunately, our medical response times, our [shark] surveillance, is getting much better, so we're also saving more [victims globally].'

The shark bite capital of the world is actually a small stretch of beach in Florida called New Smyrna – but so far none of its attacks have been fatal. In Australia, where sharks near the coast are generally bigger, encounters carry more danger. 'Comparing white shark bites to the [main] culprit species [in New Smyrna], a blacktip, is like comparing dingo bites to hamsters,' Naylor says. 'You've got about a minute-and-a-half before you faint if the big femoral artery [on the thigh] is severed.'

Leonardo Guida, a shark ecologist at the Australian Marine Conservation Society, says a perfect storm of changing environmental conditions might now be driving these big sharks even closer to the Australian coast. Warming oceans are shifting the migration patterns of some whales and fish further south – and sharks follow the food. A recent study found climate change was already seeing bull sharks move into waters around Sydney in greater numbers.

At Macquarie University, biologist Nathan Hart agrees more people in the surf does not alone explain the doubling of bites in Australian waters since the 1990s. He and his team examined more than 100 years of local attack records against environmental conditions and shark behaviour to develop a world-first predictive model for bites. 'It's a bit like weather forecasting [for sharks],' Hart says. The risk goes up near river mouths where run-off tends to draw them close, or after high rainfall stirs up the sea.

And big sharks like to hunt where cool waters meet warm. Shifts in the East Australian Current due to climate change are pushing these nutrient-rich upwellings closer to shore, Hart says. A major hotspot for white shark attacks on the east coast even lines up to the western boundary of the current. More upwellings during the La Niña event of 2020–21 could also be partly behind the latest 2020 spike, Hart says, though 'it's a bit too early to say for sure'.

Do sharks 'hunt' people?

Fifteen years ago, diver Cristina Zenato began removing fishing hooks caught in the mouths of local reef sharks in the Bahamas. Today those sharks will lay in her lap like dogs. Zenato, known to the internet as 'the shark whisperer', says people have made monsters out of sharks because it's easier to pretend they are not an animal that feels and hurts.

'We've done it to wolves and snakes, to bats,' she says. 'Three-year-olds are taught to fear sharks, even before they've seen the ocean.'

Perhaps it's because they catch us so literally out of our depth, in their domain. At Sydney's Bondi Beach, even ancient rock carvings depict a shark attack. During a brief but controversial program in 2014, the West Australian government hunted and killed 'rogue sharks' thought to have mauled people, as if tracking down wanted criminals.

Leading shark scientist Yannis Papastamatiou agrees that people still see sharks as serial killers. 'I've swum with whites, with hammerheads, I've looked into their eyes,' he says. 'They're not harmless, they don't care about your feelings, but they're not out to get you either. They have very basic reasons for attacking, which we still don't fully understand but we do know it's not malice or spite.'

While shark attack forensics are still evolving, Naylor says there's no evidence in the data so far of the same individual shark attacking more than once. Certainly, tagging research suggests sharks move around far too much to be behind numerous attacks. Even Peter Benchley, the author of the shark attack horror *Jaws*, once penned an open letter pleading with Australians not to buy into the myth of the rogue shark. 'Such creatures do not exist, despite what you might have derived from [the film],' he wrote.

Naylor says in the rare cases such maneaters have emerged in other species, say, lions and leopards in Africa, the animals were often older, losing their teeth, and no doubt beginning to see an unsuspecting human as easier prey than a baboon. But sharks, like wine, improve with age, growing bigger and more savvy.

We are within the size range of typical fare for a big shark such as a white, but Papastamatiou notes that humans have very little

*But sharks,
like wine, improve
with age, growing
bigger and
more savvy.*

fat, or energy-rich blubber, compared with some of the sharks' favourite prey, seals. 'For the same reason, you'd be able to tell if I fed you non-fat brownies instead of ones made with real butter.'

Sharks might be styled as the perfect killing machine but, as with most predators, their success rate at catching prey is surprisingly low. 'They can burn out very quickly,' Papastamatiou says. 'So, it makes sense for them to bail out if the animal doesn't offer much energy or their chances are low.'

Then why do sharks bite?

The records do reveal some crucial clues. 'The bigger [and so older] ones tend not to be involved in the incidents; they know we're not food,' Naylor says. 'It's those young, more naive sharks that are starting to transition into eating things like seals and turtles, not just fish, as they grow.'

The thugs of the shark world, bulls are jumpier than other species, with a blunt nose on a thick muscular body (and a tendency to headbutt their prey). They're tough, too, known to switch between salt oceans and freshwater rivers without skipping a beat. Their typical lifespan is 16 years.

Illustration by Jamie Brown

Papastamatiou has watched sharks hunt all over the world and, while he agrees it's usually the juvenile sharks still learning the ropes that are involved in attacks, he doesn't buy the popular theory that sharks mistake humans, often in their sleek black wetsuits, for seals.

A shark, say our white Shark 26, sees in blurry low resolution, like someone who has forgotten their glasses, but his eyes are good for a fish, and are even better at night. 'It's a bit insulting to the shark to say they're bumbling along making mistakes,' Papastamatiou says. 'Every now and then, probably, but not all the time.'

Guida agrees. Scientists (including Hart) have shown that sharks appear to be totally colourblind – able to perceive contrast and contour, but not distinguish colours – putting an end to old fears about 'yum-yum yellow' attracting them to surfboards. 'Even something as small as a cigarette packet floating on the water, sharks will check out,' Guida says.

And sharks are notoriously curious – investigation or taste testing is considered a likely motivation behind many attacks. 'They don't have hands, so how do you investigate something if you're a shark? You bite it,' Papastamatiou says. 'Of course, it can still be devastating.'

> *They don't have hands, so how do you investigate something if you're a shark? You bite it.*

This might also explain why human encounters with the predators typically look very different to a shark truly in hunting mode. When a white such as Shark 26 has spotted a target, he will barrel to the surface in a burst of speed so powerful he'll often breach out of the water himself. His eyes will roll back in his head to protect against an errant flipper or tooth.

'If sharks went after surfers the same way they went after seals we'd be seeing a lot more fatalities,' Papastamatiou says. 'Some attacks may just come down to how hungry a shark is, but very rarely do we see actual consumption.'

Many sharks will leave after biting a human rather than charging a second time or waiting for the victim to bleed out, as they often do with their regular prey.

Still, there are aberrations that haunt the public memory. The shark that mauled Ken Crew in the shallows of a Perth beach in 2000 then turned on his rescuer. Pearson says some Bite Club survivors have been bitten more than once or chased back to shore. 'We've had a shark bump someone else out of the way to go for one of our guys,' he says. 'And I've spoken to about ten people thrown metres into the air in a stealth breach attack – thankfully, the surfboards took the brunt of it.'

Still, there are aberrations that haunt the public memory.

In Tasmania in the winter of 2020, a boy was dragged from a fishing boat by a breaching white shark but rescued with only minimal injuries by his father as the shark let go. Papastamatiou says the boat itself, rather than the boy, was probably the target. 'But it's really guesswork, why it attacked.'

Hart notes many victims are surfers, who tend to be further out in the deeper water where bigger sharks dwell. But he thinks the mistaken identity theory shouldn't be discounted too quickly – his own experiments carting foam seal 'decoys' behind boats for white sharks in South Africa found that when the foam seal was outfitted with a flashy strip of lights, disguising its familiar silhouette, the sharks stayed away.

'And someone on the surface can look pretty similar to our foam seal.'

It's also unclear if sharks are territorial, Papastamatiou says. 'Some do have home ranges and . . . hierarchies. You'll see injuries from other sharks, a white might charge another one to let it know to back off, though I've never heard of [these clashes] being fatal.'

But, while some of the smaller sharks may perceive us as a threat, he says the bigger species involved in most incidents (bulls, whites and tigers) are unlikely to feel threatened by an ungainly human in their ocean.

'We need to look, too, at all the times a shark was there and nothing happened. If people knew how often there was one in the water with them, they'd be shocked.'

What have we learnt from close encounters?
Australian champion surfer Mick Fanning famously fended off a white shark live on television during a competition in South Africa, and recalled later how he felt the predator move behind him as he tried to swim for shore. Some instinct told him to whip back around, to fight with his fists, rather than turn his back.

Papastamatiou, himself trained in martial arts as well as diving, admits the old punch in the nose likely won't do a whole lot of good. 'Obviously, you should fight back, but if I had a choice I'd be going for the eyes or the gills; they're more sensitive to damage. My Brazilian jiu-jitsu probably won't do much at all unless I wanted to put the shark in a headlock – definitely not advisable.'

Instead, he says, think like Mick: 'Never turn your back on a shark.' If a shark thinks it can catch you unawares, you're a much more interesting prospect than someone vigilant who might give it a fight. Papastamatiou has seen sharks bail out of charges just because a turtle glanced up and spotted them.

Pearson shudders to imagine how he would have reacted if his own encounter with a bull shark hadn't happened so fast, if he'd looked it in the eyes as it broke out of the water in front of his surfboard. What he remembers instead was the gaping teeth and, later, a grey shape in the roiling surf. The shark's nose clocked Pearson in the head and his left arm became stuck between its jaws and his board as they were dragged underwater together, man and shark. 'I wouldn't say we were wrestling exactly,' Pearson recalls. 'We were both stunned from the [collision so] it did the hard work for me.'

Named for the stripes on their backs (which tend to fade with age), tiger sharks have a reputation as scavengers. A tiger at a Sydney aquarium famously helped solve a murder case in 1935 when it vomited a human arm with a distinctive tattoo. Tigers prefer warmer seas and can hunt in shallower water than other big sharks. They typically live for 15 years.

Illustration by Jamie Brown

In the years since that day, Pearson has come hair-raisingly close again and again to large sharks in the waters of New South Wales – 12 encounters within his first 12 months back in the water. A white 'sniffed [his] feet' all the way in on a wave. A bull rubbed its back against him.

Some close calls were easy to shake off. Some weren't. 'I'd tell [my mates], "Don't leave me alone." Sometimes I just sat back on the beach and cried, thinking, can I keep doing this? Now I'm more settled, my love of the ocean is still stronger than my fear of sharks. But they don't call me shark bait for nothing; I seem to have a radar these days. I know when to get out . . . People need to listen to sightings.'

In the 2020 documentary *Save This Shark,* Fanning himself gets close to some of the world's biggest sharks in an effort to understand what happened that day in South Africa and how the predators, which he now calls 'the janitors of the ocean', are faring against overfishing. 'I think people expected that I'd be calling for a cull on sharks, but it's the opposite,' Fanning says. 'I learnt to dive so I could get closer to the sharks and resolve the feelings I had ... Hopefully, [now] I'll be known less as "the guy who punched a shark" and more as an ocean activist.'

Shark biologist Charlie Huveneers took Fanning cage-diving with whites in the Neptune Islands, a known hotspot for the species in South Australia. It was in these waters that Huveneers and his team first noticed whites seemed to be using an interesting tactic to hunt – coming at prey from the same direction as the sun to seemingly improve their vision (and dazzle their target). They even changed direction as the sun shifted throughout the day.

'World War II jet fighter pilots did it too,' Huveneers says. 'But we hadn't seen it with a marine animal before.'

Shark senses are good, particularly their hearing, thanks to special jelly-like pores along their sides that detect vibrations. 'But that's still not going to bring them in more than a kilometre away,' Papastamatiou says. One impressive, although even more short-range sense is their ability to detect electrical fields produced by prey, say from hearts beating, allowing sharks to zero in on hiding fish. Strong electric fields can also overwhelm this sense and so repel sharks, inspiring a range of personal deterrent devices of varying effectiveness. But there's no silver bullet.

How do sharks think?

When Zenato takes tourists on reef dives in the Bahamas, she is also followed by a group of loyal 'regulars'; sharks she gives names, such as Grandma, Stumpy and Shredder. Some are more 'popular', others must be coaxed in for a feed, cheekier animals might be scolded for stealing bait. It's like the whites Papastamatiou works with in Mexico; certain animals will swim calmly to the boat, he says, 'and others are just – there's no other word for it – bad-tempered'.

Researcher Catarina Vila Pouca has seen the same surprising personalities surface in (smaller) sharks she's trained back at Macquarie University's Fish Lab. Some of the Port Jacksons on which her team ran cognition experiments were bolder than others. The sharks could also learn from one another, recognise patterns and even count (in a sense), identifying specific quantities of dots on cards.

It's likely such skills translate to bigger species, too, Vila Pouca says, as they seem key to survival. 'There's a whole range of abilities science has assumed only happened in mammals we're now testing in fish. People think fish have terrible memories, but sharks can learn things and remember them for more than a year.'

There are also signs of learning in the wild, she says. It's why fishermen will often complain of sharks stealing fish straight off their hooks or following their boats, and why tagged sharks released by scientists will often shoot off into the open ocean and not return to shore for months.

Huveneers says that personality and unpredictability can even show up in migration patterns. Sharks are constantly moving, not just to hunt but to return to preferred breeding grounds – usually in more sheltered reefs and mangroves where baby sharks born tough but small can be safe to fend for themselves. They don't travel in packs but many species have been known to come together in small 'clans' year after year.

'There's even this mysterious bit of ocean off Hawaii, in the middle of nowhere, where white sharks near the US gather at the exact same time,' says marine biologist Olaf Meynecke. 'We call it the White Shark Cafe. In the shark world, usually the bigger sharks move around more, but we still don't quite know how they find each other again in all that vast ocean. It's like a desert.'

Whites are found in every ocean on Earth. Those in Australian waters will often swim across to New Zealand, New Caledonia and the Pacific Islands. One now-famous white named Nicole (after the actress and shark enthusiast Nicole Kidman) was tracked from South Africa to south-west Australia and back again in nine months.

Studying an animal as elusive and temperamental as a white shark often makes research considerably more difficult for marine biologists than their counterparts studying species in savannas or jungles back on land. But the CSIRO has started building a white shark family tree of Australasia using DNA samples from tagged animals. 'It's like ancestry.com but for sharks,' a spokesman says. Analysing that data, they could estimate the true size of Australia's white population for the first time. During the twentieth century, numbers plunged by 90 per cent. And today, while white populations have stabilised since protections were introduced in the late 1990s, they remain in trouble, short on sharks of breeding age.

During the twentieth century,
numbers plunged by 90 per cent.

Why are sharks in trouble – and how should we live with them? Sharks have swum in our oceans for the past 450 million years – their ancestors fought the dinosaurs. They are finely evolved to hunt at the top of the food chain. Zenato calls them the wolves of the sea, stopping any one species from getting out of hand and throwing off the ecosystem below.

The problem for sharks is that they also mature and reproduce very slowly. So if their populations start to decline rapidly, as they are today under unprecedented overfishing, they cannot make up for the losses fast enough.

Naylor muses that while sharks have already survived four of the five big extinction events on Earth, and will likely survive climate change, too, 'the one thing they won't survive is being fished out of the water by a bunch of monkeys, [without] end. And there will be consequences if we remove sharks; we don't know exactly what yet. It could be algae all over our beaches, blanketing the Gold Coast. It could be much worse.'

In the Bahamas, shark numbers are stable thanks to new protections, but they still regularly find themselves snared by hooks after being drawn to fish thrashing on lines. 'It's the job of the shark to clean up what is hurting, what is bleeding,' Zenato says. And she now considers it her job to help them where she can.

Removing hooks from the mouth of a shark is not easy. 'People say, "Use pliers." I need to use my hands.'

The sharks will usually hold still for at least five or six seconds before swimming out of her hands and returning for another try.

'Some let me try over and over for 40 minutes,' she says. 'One big girl, Foggyeye, she never came into my lap. Then she got the biggest hook I've seen, I had to put my whole hand in. And she never left me alone after that.'

Sharks outside her regular group will turn up on dives with hooks, too. 'And I'll take them out and never see those sharks again. But they know to come.'

Guida has studied the impact of shark mitigation and fishing methods on animals, measuring the build-up of chemicals in their bodies during capture. 'A shark jerks just like a human when I draw blood,' he says. 'Their brains, their physiological responses to pain and stress are not that dissimilar to ours.'

It just struck me how gentle they can be, how shy they can seem.

Even in the Great Barrier Reef world heritage area, Guida says fishing nets can stretch for more than a kilometre. 'Once something's in a net, on a hook, flopped up on a boat or trawler, it becomes a race against time. Some species that need to swim to breathe, like the [endangered] hammerhead sharks, start choking fast.'

Marine biologist Lawrence Chlebeck at the Humane Society International says living with sharks safely is not about shark-proofing the ocean. 'That's a very Australian idea,' he says. '[Bites] are horrific, and they make us think authorities should do something, but the way to stay safe isn't what might feel right; it's not [vengeance].'

Research consistently shows that killing sharks – through netted beaches, baited drum lines or bullets – doesn't stop attacks. In some cases, sharks and other marine life caught, such as turtles and dolphins, might even attract more predators to the area. In others, Guida says the measures can create a false sense of security for beachgoers. 'The nets, for example, only go up four metres.' The death of a surfer on the Gold Coast in September 2020 was at a beach with both nets and drumlines, although proponents insist overall deaths remain low at netted beaches, even if attacks have not slowed.

Scientists now see real promise in shark surveillance programs being rolled out at beaches from South Africa to Western Australia and New South Wales. 'I'm saving up my pennies for my own drone,' Pearson says. 'And there's nothing like surfing with five or six survivors – one splash and every head snaps around. The big question we all wonder – why did it happen? – we never really get an answer to, so [as a group] we've become each other's answer.'

As Zenato describes the swollen jaws of the sharks she helps by hand, she stresses that she understands the risks ('I would never try to remove a hook from a great white'). But still she wonders, when the animals sink down into her lap, do they enjoy her touch?

The protective chainmail she wears over her wetsuit is soft against the skin.

In the same way we might wonder if a 500-year-old Greenland shark, born before the Industrial Revolution, has felt the ocean warming in the centuries since, as pollution spilled black and sticky into the seas.

'At the end of the day, it's a wild animal, who knows?' Guida says. 'But I remember now cage diving with a white and in the whole 45 minutes she was close, she only bared her teeth once. It just struck me how gentle they can be, how shy they can seem.'

Zenato thinks the reef sharks come to her because they feel no threat. 'Some stay for a very long time, some just a few minutes. But in that moment, they trust me.'

17

WHAT IS THE POINT OF TABLE MANNERS?

Don't put your elbows on the table, don't talk with your mouth full. Who made up these rules, and what are they for?

Felicity Lewis

It was once an article of faith, in some Australian families, that one should be fully equipped to dine with the Queen. It wasn't that the doorbell might ring and it would be Her Majesty popping in for a spot of lunch. It was more that one should know how to handle oneself – and a dizzying array of cutlery, glasses and goblets, dinner rolls and butter pats, troublesome foods and fellow diners – should one crack it for an invite to a fancy regal do.

Nowadays, dinner with the Queen is less likely to be on our minds, but table manners still matter. Attitudes to them vary, however. Adhering to them is a sign that you value 'the whole food, eating thing', says a Melbourne hairdresser whose parents migrated from Mauritius, a former French and British colony. They are a way to show respect, particularly for one's elders, says a chef who grew up in Malaysia. They help to build relationships, says an etiquette expert in the United States. They can reflect an *Upstairs, Downstairs* morality designed by the elite so they remain 'the cherry at the top of the tree', says a Catholic priest who grew up poor in Melbourne.

And yet every family follows table manners in its own way, from those who pepper their urbanity with the odd broken rule – 'Whoops, I may have just passed the port to the *right*!' – to those whose impulses override etiquette – 'I totally [*crunch, crunch, crunch*] disagree!' Even in families where no one mentions elbows, there are always behaviours at play when sharing meals. And there are many common threads to the rules, even as differences in table etiquette across cultures have long vexed diplomats, traders, travellers and other citizens of the world.

So, what are considered 'good' table manners? Says who? And why can't you put your elbows on the table?

Where did table manners come from?

The custom of families meeting for meals goes back two million years 'to the daily return of protohominid hunters and foragers to divide food up with their fellows', writes Margaret Visser in her fascinating classic *The Rituals of Dinner* (1992). From her home in the south of France, Visser says, 'I start the book by saying there's no such thing as a society with no table manners. And that's why I started with cannibals, because even they have table manners – very strict ones that make a big difference between eating an animal and eating a person.' Table manners express 'all kinds of usually unconscious prejudices', she says. 'You can find out a huge amount about any society by watching them eat: who's higher than you, who's missed out, who's not invited.'

Some of the rules are codified. The *Book of Rites*, a group of texts attributed to Confucius, declares that mealtimes separate savagery from civilisation, writes Jonathan Clements in his intriguing story of Chinese food, *The Emperor's Feast* (2021). Clements quotes the ancient book to illustrate what being 'civilised' might have looked like in the fifth century BC: 'Do not roll the rice into a ball; do not bolt down various dishes; do not swill down [the soup] . . .'

> *You can find out a huge amount about*
> *any society by watching them eat.*

Centuries of Islamic dining etiquette were drawn on by Muhammad Badr al-Din al-Ghazzi of Damascus in his sixteenth-century *Table Manners*, notes University of Cambridge historian Helen Pfeifer in her article 'The Gulper and the Slurper: A Lexicon of Mistakes to Avoid While Eating with Ottoman Gentlemen' (2020). Ghazzi, she says, warns against dining types such as the annihilator (*al-mukharrib*) who leaves 'only scattered bones in his wake', the trickster (*al-muhtaal*) who slyly piles meat on his neighbour's plate and then eats it all when his neighbour politely refuses, and – shudder – 'the one who leaves greasy traces' (*al-mudassim*).

Sociologist Norbert Elias puts a thousand years of European manners under the microscope in his 1939 study *The Civilizing Process*, studded with gems that make us chortle only because we modern adult diners simply *know* what kinds of behaviour are beyond the pale. The thirteenth-century German poet Tannhauser offers this, for example: 'It is not decent to poke your fingers into your ears or eyes as some people do or to pick your nose while eating. These three habits are bad.'

By the thirteenth century, courtesy (how to behave in court) was gaining currency with a warrior nobility in Europe, writes Elias. The kingdom of Provence and the city-states and principalities now known as Italy were trendsetters (the Muslim rulers of al-Andalus, from 711 until the late 1400s, were no slouches when it came to refined courtly dining, either). The English caught on and by 1392 poet Geoffrey Chaucer was poking fun at 'curtesy' in *The Canterbury Tales*. We meet a nun whose 'upper lip was always wiped so clean/That on her cup no speck or spot was seen/Of grease, when she had drunk her draught of wine.' When merchant and diplomat William Caxton set up a newfangled printing press in England in 1476, it was no surprise that a book of manners was among the first titles he cranked out.

Caxton's *Book of Curtesye* (1477) speaks unabashedly of belching and farting at the table – 'Beware no breath from you rebounde' – as does Erasmus of Rotterdam's *On Civility in Children* (1530), which warns that fidgeting in your chair gives the impression you are trying to squeeze out a fart. Such talk of bodily functions is typically medieval in its directness, notes Elias. Life was a visceral affair. If you wanted to be delicate, you used three fingers to pick up your meat and you refrained from offering a half-eaten hunk to someone else, even if you liked them. But the nuanced advice of Erasmus, in particular, hinted at a change in the wind – the *impression* you made mattered. Power was shifting from feudal lords to a new kind of aristocracy for whom delicacy and *civilité* were at a premium. 'Not abruptly but very gradually the code of behaviour became stricter,' Elias contends, 'the sense of what to do and what not to do in order not to offend or shock others became subtler.' It grew easier to feel shame and embarrassment.

By 1605, King Henri III was being satirised for chasing peas around his plate with a pretentious implement called a fork, writes Visser, and in the lavish royal court of Versailles under Louis XIV, florid displays of feasting were *de rigueur*. It's thought the word 'etiquette' (ticket) came from place cards, which indicated where each guest was to sit at banquets. While a hereditary title could get you a place at the table, soon enough money could buy a way in, too. Learning the rules was a high-stakes enterprise for the bourgeoisie – one didn't want to commit a faux pas (false step). Culinary historian Professor Barbara Santich at the University of Adelaide has pored over the texts on table manners from these times. 'Very often when books of etiquette are done, it's to enable people to improve their social situation,' she says. 'After the revolution in France in 1789, the "father of food writing", Grimod de la Reynière, wrote a number of books about how to eat in a restaurant and how to entertain at home and he said, precisely, it is for the nouveau riche who have flooded into the capital.'

The peak of fusty formal dining may well have been in the nineteenth century, says Santich. It was by then accepted that cutlery was a good idea, but the Industrial Revolution meant factories could pump out the stuff. Where once you had to be born with a silver spoon in your mouth, now a middle-class person could afford a whole set. 'You had a fish knife, an oyster fork, a cheese knife,' says Santich. 'You would have a teaspoon and a coffee spoon and a soup spoon and dessert spoon.' Crockery proliferated too. 'There were tea cups and little coffee cups and saucers . . . There was a vegetable dish, there was a fruit plate . . . A dinner service might have a dozen different items for each person. For each item of cutlery, there had to be a new rule.'

But in the colonies, manners were more relaxed, right?

'I was brought up to have table manners,' says celebrated chef and author Tony Tan, who grew up in coastal Kuantan in Malaysia, eating Indian, Chinese and Malay cuisines with chopsticks, hands, spoon and fork. The Federation of Malaya became independent of the British in 1957. Tan's parents ran rest houses for the British where his mother cooked roast chicken and trifle. One of Tan's earliest memories is of watching Indian road workers eat lunch. 'They unbundled their bag of food. They were eating with their fingers, and I was salivating.' Seeing the little boy looking peckish, a woman rolled some rice and curry into a ball and flicked it deftly into his mouth. 'I burst into tears because it was so hot, chilli hot . . . It was like the pain and the ecstasy of it all: too hot to eat but so beautiful to swallow.'

It was into this pungent cultural mix that a Mrs Windsor (no connection to the Queen) arrived to instil 'Britishness' into 'us natives'. 'All I can remember was very heavy, red velvet curtains and all the cutlery was being laid out on the table,' says Tan. 'What is a fork? Knife? Serving knife? All those things that put the fear of God into all of us. And then we've got to start eating, from the fish knife to the oyster fork. And that was really very daunting, particularly for an eight- or nine-year-old who'd never actually ever eaten an oyster in his life – those horrible, squiggly-looking things! And I was just thinking, why is she wearing stockings, because they are just so hot?'

Tan, who went on to train as a chef in Paris and London, is an expert in Asian cuisines from Cantonese to Malaysian, which he teaches at his school in country Victoria – but the etiquette, particularly of his Chinese heritage, has remained. He says, 'You've got to invite your elders to start eating, or say, "We are now eating" so the elders can say, "Go ahead." It's a sign of respect to people who are older than you. And so, when people don't do that any more, they lack manners and they lack good upbringing.'

In a socially mobile colony, it is manners more than a family coat of arms that 'reveal to us the lady and the gentleman', declares the *Australian Etiquette, or the Rules and Usages of the Best Society in the Australasian Colonies* in 1885. 'Manners and morals are indissolubly allied,' it contends, 'and no society can be good where they are bad.' Naturally, it is 'the duty of Australian women' to ensure the development of this moral fibre, with a view to Australia becoming 'the best society of any country'. Colonists are advised to practise their table etiquette at home, even when eating alone, lest they become 'stiff and awkward' when out. Among the many little points to be observed: 'If anything unpleasant is found in the food, such as a hair in the bread or a fly in the coffee, remove it without remark.' (It's hard to imagine anyone today keeping quiet about a fly in their macchiato.)

As Barbara Santich points out, 'Sometimes you've got to look at the books as trying to correct a situation, not necessarily reflecting [it].' Free from the strictures of British deportment, colonials did relax some of the rules. 'The picnic became terribly, terribly popular in Australia,' she says, 'much more so than in England. The weather had something to do with it but it was also symptomatic of an attitude: we can be a little bit more free and easy, and possibly egalitarian – we can do our own thing. The picnic was, in a way, a deliberate infringement of table manners.'

Barbecues do away with some of the rules, too, says Visser. 'Even having a table means we choose who we're going to feed, so the barbie is a wonderful way of breaking that down.' They also dissolve the hierarchy that comes with sitting at a table, although she suspects not entirely, noting how it tends to be the men ' doing the fire'.

Today in Australia, says Santich, 'You look at people in restaurants. There are ways that some people hold their knife and fork that would have been frowned on, or even a more American style [where diners cut one piece of food then put the knife down, transfer an upturned fork to the knife hand, and use it to bring

The picnic was, in a way, a deliberate infringement of table manners.

the cut morsel to their mouth] – you wouldn't have been allowed to do that. Other manners have come into our society and been accepted and incorporated just as other foods and dishes have been accepted and incorporated into what we do, what we eat.'

Surely, in the United States – the land of the free since they declared independence from the British in 1776 – table manners are more relaxed? Pamela Eyring, who heads The Protocol School of Washington, says America is like Australia, insofar as 'we are more relaxed and more casual people'. As with anywhere, table manners in the States are all about context. Take hamburgers. 'When I was the chief of protocol at Wright-Patterson Air Force Base in Ohio, we had German military counterparts who were visiting, and they wanted to go to an authentic cheeseburger restaurant. So I took them and they *had* to use a fork and knife. I said, "Culturally, you have to pick it up and bite it – it tastes better this way,"' she laughs. In upscale restaurants, though, she uses cutlery to eat her burger: 'A lot of people are watching their weight or reducing their carbs or sugars or fats and so what we're finding is that they take the top bun off and place it to the side and then use the fork and knife.'

The allure of the old world is sent up in a 1994 episode of the TV comedy *Seinfeld* when a trend catches on for eating candy bars with cutlery. 'Forgive me for trying to class up this place, for trying to have the Yankees reach another strata of society that might not watch Channel 11,' says George Costanza to colleagues at the New York Yankees, as he slices into his chocolate bar on a plate. A co-worker deadpans, 'What the hell are you doing?' Costanza effects a debonair swagger as he waves his forkful of chocolate in the air: 'I am eating my dessert. How do you eat it – with your *hands*?'

What is behind table manners?

'A meal is always both love and violence,' says Visser. 'When I gave lectures on this, [people would say] "There's no violence at the table, what are you talking about?"' And yet there is a distinct possibility of violence 'when you're sitting around with knives and forks and you're all hungry,' says Visser. 'Wow, if ever there's a place for violence, there it is!'

Rupert Wesson, a director at Debrett's – an authority on how to behave and who's who among the peerage of the United Kingdom since 1769 – nominates avoiding violence as an ancient driver of etiquette. 'If you actually sat down for a meal, to a certain extent you were vulnerable. Was the food poisoned? Would that knife I'd just given a stranger end up being plunged into my chest? So [today], it's rude to wave your knife, point your knife. The knife should never really come up much above the height of the plate.'

As Tony Tan notes, in Chinese cuisine 'everything is cut up into small pieces so there's no need to use a knife'. (It's also considered rude to point at someone with chopsticks.) In *The Emperor's Feast*, Clements recounts a banquet at Swan Goose Gate in 206 BC to which an army general invited his great rival; but things went pear-shaped when a soldier was invited to perform a 'sword dance' for the guest, who was bundled away soon after.

That knives give us the jitters explains a few things: why the cutting edge must face the plate in a setting; why we tend to wince if someone licks their knife; why we are taught not to hold cutlery in our fists, like a weapon; and why Claude Calviac's advice, in *Civilité* in 1560, still stands – 'If you pass someone a knife, take the point in your hand, and offer him the handle, for it would not be polite to do otherwise.' It's also why table knives are so blunt, evolving from spikes to double-edged blades to . . . rounded off. Only steak knives have survived to be sharp and pointy.

There are more positive reasons for table manners, too. In the Punjabi village where chef and restaurateur Jessi Singh grew up, families ate together sitting on jute mats on the floor. There was no electricity or running water. Some of his friends did not have enough to eat. 'The biggest manners would be, you can't leave anything and you eat happily whatever you get in front of you.' The four fingers of the right hand create a cup and, using the bent thumb, 'you just *slowly* put food in your mouth without opening your mouth too much'. 'Food is such a sacred thing in India, when you touch food and it goes in your mouth, you bring all those positive elements of your life: you work so hard to get there that you are able to eat.' It's little wonder Singh is troubled by the dishes left unfinished in his restaurants in the United States and Australia 'and all the food that goes in waste'.

Grasping at food is just not the done thing. 'If you think about something really simple such as reaching across the table,' says Wesson, 'and everyone says, "Gosh, that's rude, we don't do that!", you can even run that back [in history]: that's the idea of grabbing more than your share.' As Erasmus observed tartly in 1530: 'Some people put their hands in the dishes the moment they have sat down. Wolves do that.' Tan watches in horror when diners pile portions of every single dish on offer into their bowl of rice in one go. Diners should pick 'very gently and politely' at dishes on a communal table, 'and if you've touched that piece of parson's nose [on a chicken], you've got no choice but to pick it up and put it on to your plate'.

Slouching is poor form, too, thus 'Sit up straight!' Every place at a table has a boundary made up of the cutlery and the space between chairs, says Visser. Keeping your elbows tucked in, and not spread out on the table, is a way of not invading your neighbour's space (which might, in the end, cause violence). An exception, she says, is the diner who places their elbow with an 'elegant lightness' that makes it clear they are not supporting themselves on the table and don't need to do so, and who has shown in everything else they do that they 'have *earned* this nonchalance'.

Before he coached etiquette, Wesson was an officer in the British Army, suiting up for regimental dinners in barracks 'with all the sort of stiff formalities that you might associate with British etiquette' but mucking in for more rustic meals in Bosnia, Iraq and Sierra Leone. During his deployment in West Africa, in a team of 25 soldiers from 17 nations, meals were 'absolutely the thing that bonded us together'. At times, though, especially with memories fresh of the deadly Ebola virus, and other bugs still afoot, the role of hygiene in driving etiquette was heightened. 'You would eat in a way that was not messy. You wouldn't use your knife and fork in the communal bowl.' In India, hand-washing is a given. 'You're using the right hand to eat food because your left hand you use for your bum,' says Singh. 'In India and most parts of the world, we still use squat toilets and you use running water to clean yourself.' Hygiene is behind many of the manners Santich studied from fourteenth- and fifteenth-century Europe, which were not unlike the ones drummed into her as a child: 'Not putting too much in your mouth, not talking with your mouth full and not wiping your hands on the tablecloth. The three of them I would call simple hygiene that would arouse disgust if you did them, and therefore you don't.'

Film director Quentin Tarantino, no stranger to gore, reportedly said the only film scene he ever found truly disturbing was the Mr Creosote sketch in Monty Python's 1983 *The Meaning of Life*. In it, a projectile-vomiting glutton played by Terry Jones is attended to by an obsequious maitre d' played by John Cleese ('But it's wafer thin'). The provocations to disgust are all there: creosote is actually a stinky liquid painted on wood to stop it rotting; and when Creosote requests every appetiser on the menu all mixed up in a bucket with 'the eggs on top', it's hard to know which eggs he means – the eggs Benedict, the Beluga caviar, the little quail's eggs . . . 'Our cultures hate slime,' says Visser. 'The problem with it is that it does not keep to our categories that separate the hard from the soft, it combines them.' Nobody wants to be a Mr Creosote, nor to sit near one, not anywhere. In India, says Singh, when 'your mouth is full of rice and, while you're trying to chew, all the food

is coming back on your plate, people are like, "Oh, my goodness."'
Tan remains incredulous when someone spits food particles on
the table while speaking: 'Why can't you just swallow your food
first and then, afterwards, you know, start talking?'

Can you have bad table manners and succeed?

Most senior US military leaders have good table manners, says
Pamela Eyring, but not everyone grows up being taught the finer
points of formal dining. On attaining the rank of brigadier-general
or senior executive service positions, air force leaders are sent to
what used to be called charm school, 'and so many say, "I wish I
would have had this earlier in my career."'

Eyring, who started her working life as a stenographer at
Ohio's Wright-Patterson Air Force Base – 'I couldn't spell the
word protocol' – was its chief of protocol by the time she left
23 years later. She rolled out red carpets for visiting presidents,
including Bill Clinton. Any head of state will have a team to advise
on etiquette, she says, but 'they have gaps, they're humans . . .
they forget, or get casual. I mean, you might get sick.' George H.W.
Bush famously threw up in the lap of Japanese prime minister
Kiichi Miyazawa while suffering from gastro during a state dinner
in 1992. 'Roll me under the table until the dinner's over,' he
reportedly said to his doctor while he was on the floor.

Thomas Jefferson's table manners caused a furore that
ultimately enhanced his reputation when he hosted a White House
dinner in 1804 at which the brass-buttoned Brit Anthony Merry
and his wife, Elizabeth, were guests. When it came time to be
seated, Jefferson, a widower, escorted one of his minister's wives
to the table, as was his custom. A shocked Elizabeth, expecting
top billing on the president's arm, was instead escorted by the
secretary of state – to a seat that was not beside Jefferson. Her
fuming husband had to find his own spot. Merry thundered to his
bosses in London about this 'absolute omission of all distinction'

so Jefferson, despite his distaste for such concerns, instituted a 'pell-mell' rule, where everyone sat wherever they could find a place. 'He wasn't rank-centric and believed in a more egalitarian approach to social and diplomatic relations,' says Eyring. 'He wanted it to be like, "I'm not special, I'm not the king, I just want democracy."'

Career diplomats, though, had better get their seating right. 'Meals are a tool of the trade both in conveying messages and forming relationships,' says Richard Rigby. As a diplomat from 1975 until 2001 in Tokyo, Beijing, Shanghai, London and Tel Aviv, he researched etiquette before each posting but learnt 'tricks for young players' on the job. He recalls a Chinese vice-minister rushing around tables at a dinner at the Australian ambassador's residence in Beijing just moments before guests entered, saying, 'No, no!' as he switched place names. Although they had planned the occasion 'by the book', the Australians seemed to have erred in seating their guests by ministerial title rather than Communist Party rank.

Sometimes, the meal is the message. When then opposition leader John Hewson visited Beijing in 1990, one of the first Western leaders to do so after Chinese troops killed protesters in Tiananmen Square, he declined to attend the Asian Games 'to avoid being the focus of propaganda'. He was duly given the bum's rush. 'The warmth of the initial introductions faded as the meetings proceeded,' he later wrote, 'to the point where, at the last formal banquet, a 13-course meal was served in about 12 minutes.' Rigby, who was there, too, chuckles at the memory. 'Normally, you would have an hour and a half. Courses were whipped away before you had a chance to get into them. I've never witnessed anything quite like it.'

As part of his job, Rigby ate sheep's eyes and camel hump – 'sometimes it's just a test to see if you can' – and had his drinking limits tested thoroughly, particularly in China where 'death by drinking' is a tradition. Feigning a heart condition, saying you are a teetotaller or, if desperate, emptying your glass of clear spirits under the table were all diplomatic ways out. Rigby, who is now an emeritus professor at the Australian National University's Centre on China in the World, notes that leader Xi Jinping's anti-corruption push has meant more modest banquets and less 'weaponised drinking' – a 'great relief' for diplomats.

In business, table manners often come into play in cross-cultural situations. Eyring has offices in Dubai and the US capital, and graduates of her school's certificate programs in 90 countries. Her clients might hire the school to train their staff on intercultural etiquette and cross-cultural communication in order to build their business or, say, if they are from the UAE, to understand Western customs and do business more effectively with international companies. It's all about 'how you communicate with people to build relationships,' she says.

Wesson criss-crosses the world (when there is not a pandemic on), working with senior executives preparing to be parachuted into overseas roles. He also trains fresh-faced graduates, encouraging them to engage, smile, make eye contact – 'all the things your parents always said you should do'. He says learning etiquette can become a vehicle for improving self-confidence. And the Brits are actually a fairly rules-averse bunch. 'We like having rules and guidelines but only so that we can break them. You have to understand what the guideline is, and know that you're not following it – and to know that you're not following it with a certain amount of insouciance. That is absolutely fine.'

So, what is the point of table manners today?

'There's etiquette and there's etiquette,' says Father Bob Maguire, a priest since 1960. 'The best etiquette is to make sure that you put other people first.' Just as sit-down meals can exclude people, they can also welcome people in. Maguire hosts a big Christmas lunch in Melbourne every year. 'We have 100 people sitting there with knives and forks and plates and God knows what they haven't seen before, but they do their best, you see,' he says. 'The best manners is to make sure that other people feel comfortable in your presence.'

> *'Etiquette is more about care and consideration, and all of those things that allow everyone to sit down at the table as equals, to share food and to feel comfortable doing it.'*

What would Maguire say to an etiquette snob? 'If people are sitting around a table and noticing that your mother is eating her peas with a knife, and if they want to have a go at you, your mother or whoever it is that's sitting there doing what she's been doing for the last God-knows-how-many years because they were born poor . . . well, what do you say to them? "Behave yourself – that's not good manners, to criticise!"'

There's an apocryphal story, says Wesson, about the Queen (in most accounts, Queen Victoria) entertaining guests and everyone being served prawns, which, according to etiquette, are broken open with the hands; the diner then rinses their fingers in a little bowl of warm water. But at this meal, a foreign dignitary picked up his finger bowl and drank from it. The Queen, without missing a beat, picked up hers and drank from it, too. 'Etiquette is more about care and consideration,' says Wesson, 'and all of those things that allow everyone to sit down at the table as equals, to share food and to feel comfortable doing it.'

He points to a 'weaponising' of etiquette where people are 'consciously and overtly' judged for having 'incorrect' table manners. 'The sort of people who would notice and make a mental note probably aren't the sort of people you want to be dining with anyway,' he says. And invoking 'novelty points' such as how to eat asparagus when it is served on its own (answer: with your fingers) is just showing off: 'Look what I know about asparagus!' If there's any trick to holding your own at a formal do, it's to be so interested in and interesting to the people around you that no one thinks about table manners, he says.

It was as a student attending the protocol school she now heads that Eyring had an 'aha' moment and realised the detail she hadn't been aware of. 'Maybe I didn't find it in the books or in a culture-gram,' she says. There is, for example, 'the kindness and appropriateness of a toast, how you hold the glass, even, and how you show appreciation. It wasn't about the drinking of the alcohol. It was the toast itself and the honour of that. It's the emotions that come from protocol and etiquette – when you meet someone and they're focused on you, your company or country, and showing respect.'

Another good reason for table manners, says Visser, is that they are a kind of false morality. 'You're behaving as though you are virtuous. It's not real virtue but if you pretend to be virtuous, then virtue might occur.' Even people who behave very badly must at least behave themselves at lunch, she laughs. 'You can't get out of it, in other words.'

In this way, etiquette sets the scene for something that is a real virtue, that we all care about deeply. 'The reason why you don't want the violence is because you want the love. If you think about it, every single religion, without exception, is based on a sacred meal. If you're eating together, it's love. You're sharing food. You have consideration for others. You pass the mustard. You don't shout and jostle and scream. All the things that table manners are supposed to do, it's because the rule symbolises love.'

18

HOW DO YOU BUILD THE PERFECT SWIMMER?

Power dives and snappy tumble turns –
elite swimmers work hard to hone
their 'skills', and then there's their stroke.
What are their secrets?

Phil Lutton

The high-powered contests in the pool at Tokyo in 2021 didn't bear much resemblance to the rustic swimming in the first modern Olympic Games, in Athens in 1896. Then, all races were completed in a single day in a heavy swell off the coast of Piraeus. Tokyo's was a nine-day affair in a sleek, new pool with geothermally heated water and a roof design inspired by origami.

Between the Athens and Tokyo events, there has been more than a century of evolution in swimming – in stroke, science, biomechanics, training, technique and race craft – split among the four strokes of Olympic competition: freestyle, butterfly, backstroke and breaststroke. Now, every microsecond is accounted for, every angle of every manoeuvre put under the microscope. The fastest way to get from A to B remains freestyle, the ubiquitous 'front crawl', also known as 'Australian crawl' (it also appeared on carvings in ancient Egypt). Freestyle is an event that, technically, offers an athlete freedom to swim in any way they choose, such that, in theory, a competitor could swim backstroke in a freestyle event should they be so bold.

While modern front crawl can get pretty fancy below the waterline for elite swimmers, the thousands of Australians who do laps for exercise and relaxation every week are unlikely to fret too much about the finer points. And yet, when you see an elite athlete glide through the water, it's hard not to wonder: how do they do it with such fluency and speed?

Who's eligible to be a star swimmer?

Elite swimming has, historically, been a sport in which success has arrived in all shapes and sizes, yet slowly but surely, it has started to become a land of the giants. Walk around a pool deck when top performers are training and you are likely to be looking up. Expect to see some common traits, too; V-shaped torsos, broad chests and shoulders, ripped cores and powerful backs and upper arms.

It's not impossible to make it as a shorter swimmer but the

statistics and science suggest the longer, the better. At the 2016 Olympic Games in Rio, the average height of a male finalist was 188 centimetres (6'2) and for women, 175 centimetres (5'9). Some take it to extremes, with Sun Yang, the now banned Chinese star, measuring a towering 198 centimetres (6'5). His great rival, Australian Olympic gold medallist Mack Horton, is 1.9 metres (6'2) while Australia's Cate Campbell is one of the tallest elite female swimmers at 186 centimetres (6'1).

It's not impossible to make it as a shorter swimmer but the statistics and science suggest the longer, the better.

The links between height and speed in the water were outlined in a research paper by Per-Ludvik Kjendlie and Robert Stallman from the Norwegian School of Sport Sciences in 2011. 'A taller swimmer will generally swim faster,' they write. 'Larger propelling sizes [bigger feet, hands, longer arms] better the propelling efficiency and lower the stroke rate – creating a more energy-efficient mode of swimming.

'Drag is influenced by size: a taller swimmer will create less wave resistance at the same speed, and the tall swimmer will have a greater potential for maximal velocity due to a higher hull speed. During a swimming race, a taller swimmer will have a shorter true race distance due to turning and finishing actions with their centre of mass further from the pool wall.'

The by-products of height are highly useful. Larger hands can hold more water through the stroke while big feet act as propellers, as in the case of Ian Thorpe's famous size 17s, which thundered him past Gary Hall Jr in the unforgettable 4 × 100 metre freestyle relay at the Sydney Games in 2000.

But being tall can have its drawbacks. Campbell, for example, gains so much from her long arms and fluid stroke but sacrifices some explosiveness off the blocks as she hauls her rangy frame into the pool. Her dives have been a major focus of her training, for the 50 metres especially, and she has enlisted the help of biomechanics experts to search for improvement and to help activate her fast-twitch muscle fibres.

'I've really put a lot of time and effort into bringing them up to scratch. And 50s are won and lost by hundredths of a second so even if I can get off the blocks a few hundredths faster then that's a win. Improvement comes down to a lot of technique,' Campbell says.

What's in a dive?

The FINA World Championships in Gwangju in South Korea in 2019 were as dramatic as any on record, headlined by Mack Horton's podium protest following positive doping results from Sun Yang. In the water there was some scintillating racing, including a 100-metre freestyle final that perfectly illustrated how the technical skills of a swimmer can trump the pure racing speed of a rival.

Kyle Chalmers won Olympic gold for Australia in the same event in Rio when he was just 17. He is a juggernaut in the water, starting slowly over the first 50 metres and descending on his rivals as if they are treading water to unleash a booming finish. On the other hand, his American adversary Caeleb Dressel waits for no man, sprinting out of the gate and daring his rivals to catch him.

It would turn out to be an electrifying contest, with Dressel winning in a blistering 46.96 seconds, holding off Chalmers who sprouted wings late to take silver in 47.08 seconds. Dressel's time would be the fastest ever recorded by a swimmer over 100 metres in a 'textile' swimming suit (as opposed to the 'supersuits' of the mid-2000s made out of material that trapped air, increased buoyancy, and sent world records tumbling before they were banned at the start of 2010).

And yet, curiously, the breakdown of that race shows Chalmers to be the faster swimmer. When he was up and stroking, Chalmers was moving at 2.02 metres per second compared to Dressel's average of 1.99 metres per second. So how did Dressel win?

Old videos of swim races, show competitors simply flopping into the water when the starting gun goes off. The dive is a means to start swimming. Now, over the sprint distances in particular, the dive is a weapon. And it doesn't end with a swimmer leaving the blocks – it begins a phase of the race that continues with an underwater 'streamline' then a 'breakout', as swimmers spear through the water with as little resistance as possible to break the surface and start their stroke proper.

> *Now, over the sprint distances in particular,*
> *the dive is a weapon.*

Starts (as well as turns, which we'll get to) are where a swimmer is travelling at their highest speed, and Dressel is an expert. In 2019 in Gwangju, when the starter buzzer sounded, Dressel was off the blocks in 0.61 seconds, compared to 0.71 seconds for Chalmers. That's 0.10 seconds the American already has in the bank as he moves to a part of the race in which he has a clear advantage over his Australian rival.

Having hit the water, the athletes next streamline, tucking their heads downwards, stretching out their arms and overlapping their hands as they propel themselves with a series of underwater dolphin kicks. The goal here is to reduce drag and explode back to the surface with as much velocity as they can conjure. Look again at the numbers: Dressel is under water for a total of 4.48 seconds and surfaces after 13.2 metres, while Chalmers is under water for 4.28 seconds and surfaces at the 12.35 metre mark.

When we see them getting to work on top of the water, Dressel already has almost an entire body length on Chalmers as they charge to the wall to prepare for their tumble turns.

Since this race, it is the dive that Chalmers has spent hours working on. 'If I can improve my dive, within a fractional margin, I know my swim speed is my strength. At world champs, it was the reaction time off the blocks. I know that I can improve that [dive] . . . skills are something to work on day in, day out and I know they can get better.'

Why do tumble turns matter?

Turns are an underwater technique used by swimmers to change direction once they reach the end of the pool. For social or semi-serious recreational swimmers, it's a handy skill that allows more fluency in the water and avoids the 'stop-start' motion at the end of each lap. For high-level swimmers, across all distances, it's a vital part of racing that saves time and creates speed as they launch themselves off the wall.

The turn effectively starts as a half-somersault in the water, with legs tucked tight to the chest, before the swimmer extends out horizontally near the wall and outstretches the arms into the overlapped streamline position, as you would after a dive. The legs drive the swimmer off the wall, followed by a series of butterfly kicks through the breakout and back into the stroke.

It looks graceful when done correctly but, under the water, there's also a lot of power being generated that sees swimmers surge into the new lap. Over a longer race, that can add up to serious time and it's not unusual to see good turners hit the wall behind a rival and pop up ahead of them.

Alas, for those attempting to master it at the local pool, it's a frustrating endeavour that requires patience, persistence, good control and balance in the water and above all, hours of practice.

Elite swimmers loosely group manoeuvres such as dives and turns into the genre of 'skills'. The rest is called, well, swimming. But while all aspects of training are finely calibrated, skills have increasingly become critical.

'When it boils down to it, Kyle is a better swimmer then Dressel but Dressel is slightly more athletic and explosive in those movements,' says James Magnussen, a world champion 100-metre freestyler and the silver medallist in London 2012. 'He is able to control the tempo of the race because of his brilliant technical skills, which are the starts and turns. He can toy with the opposition because his skills are so much faster.'

So let's return to Gwangju in 2019. Chalmers and Dressel are churning towards the halfway mark. Here comes Chalmers. Over the remainder of the lap, he's swimming faster than Dressel, slowly reeling in some of the early gap. But Dressel is about to produce a brilliant turn and launch off the wall to give himself even more breathing room to start the second 50 metres of the race. This time he breaks out at the 9.5-metre mark after moving at 2.43 metres per second under the water. Chalmers is no slouch but with an underwater speed of 2.35 metres per second when he returns to our view, 8.55 metres into the second lap, he must resume his chase anew.

From the 65-metre mark, Chalmers is swimming faster than Dressel at every checkpoint but Dressel finishes first and is crowned the world champion.

At the Tokyo Olympics, all of the work on skills and dives and turns almost paid off for the Australian. He sped home in the final 50 metres to touch the wall in 47.08 seconds, having executed an almost perfect race. Yet there was Dressel again, the victor in 47.02 seconds, a new Olympic record. His sublime skills and athletic gifts had carried the day once more.

What about the swimming part?

Let's focus here on freestyle, the stroke most swum in pools around the world. Even for the pros, it's a constant work in progress and it's common to see even world and Olympic champions being pulled up on matters of technique by coaches. From the angle of the arm entering the water to body position to the kick – whether a waltz-like, six-beat one-two-three rhythm or, say, two kicks per stroke cycle – it's a multi-faceted movement that must work in unison to achieve the optimum result.

Swimmers might focus on a few key areas, say, catching or hooking the water with the hand – the more water you catch, the more efficiently you will be able to haul yourself forward. But breathing technique is at the heart of it all – legendary US Olympic coach Bob Bowman, who was mentor of pool great Michael Phelps, says the head shouldn't pivot independently of the body. 'Rotate your head in line with your body,' he tells US swim bible *SwimSwan*. 'Try to take your breath with one goggle still in the water.'

Other areas include body position (head down, hips up) and the idea of counting the number of strokes you take per lap, just like a runner might count the cadence of their steps. Alex Popov, the great Russian Olympic freestyle champion, believes that's important to ensure swimmers maintain their form under fatigue. When you run, the tendency when tired is to shorten your stride and slouch. Swimming is the same – your arms collapse, your stroke weakens. By counting, you measure how many strokes it takes to get from one end of the pool to the other and you try to maintain that, even when it's tough.

What actually is butterfly?

Nothing says, 'I can swim' like pounding out laps of butterfly in the pool. A confounding stroke to the uninitiated, it's actually a hybrid that had its genesis in breaststroke – inventive swimmers found it quicker to return their hands to the front position above the water, not underneath. It first appeared in the Olympics in Melbourne in 1956 and is known as the most challenging stroke due to the immense physical exertion and precise timing needed to make it all sing.

Each butterfly 'stroke' is, essentially, two dolphin kicks followed by a sweeping simultaneous arm motion that stretches forward and straight and allows the swimmer to then drag themselves along underneath the water. It gives the impression of swimmers leaping out of the water (arms and chest spread wide, roughly in the shape of a butterfly) and requires, among other things, good flexibility through the chest and upper back, a high degree of upper body strength and a powerful dolphin kick that drives the entire motion.

When executed at a high level, it's a thing of beauty. Think of the Australian swimming great Susie O'Neill, or America's Phelps, and you know what we are talking about. For the rest of us, it's more of an expert move that might be best tackled once you are fit and firing in your other strokes.

But for those determined to master it, or even those who want to just feel fluent in the water, there's a simple shortcut, says Magnussen. It's called 'getting lessons' and he believes it can produce sharp results in not much time.

'Most pools will have an adult squad and it is definitely worth doing just one a week even, just to pick up some added technique. It's the same as golf . . . everyone seems embarrassed to have golf lessons but everyone needs golf lessons,' he says.

'I've done lessons with corporate groups and within an hour, I can drop a couple of seconds off their 50-metre time. It can be so easy and obvious. You might not find someone of my experience at every pool you go to but any good coach can give you tips that will make swimming a whole lot more enjoyable.'

HOW TO LOOK LIKE A PRO AT THE POOL

For those who don't have the chance for hands-on coaching, here's Ben Allen, the director of aquatics at Brisbane's Anglican Church Grammar School, with three key areas to work on next time you're in the water.

Breathing

'Many swimmers neglect thinking about how they breathe in the water and hold their breath. What we want to do is ensure you are blowing your air out under the water in order to have maximum oxygen intake when you take your breath. This helps create a natural breathing motion that you would have on land when walking or running.

'For freestyle, I would recommend trying to breath in a pattern of two strokes to a breath or three strokes to a breath. Once you have become comfortable with a pattern, you can mix things up and alternate two strokes and three strokes to a breath on either side. Breathing to both sides is a very effective way to balance out your stroke and have more control in the water.'

Technique

'Technique is about being efficient, relaxed and holding momentum. Good swimmers swim like they are taller than they really are, reaching forward on the hand entry and extending the hand past the hip on exit. They are relaxed in their movements, making sure they aren't overly stiff and tense in their muscles, with their eyes looking down, hips on the surface to carry momentum. Being flat in the water is our number one priority; the less resistance, the easier it is to swim.

'Don't be afraid of using tools such as fins and pull buoys to help achieve these points. Swimming 30 minutes well with a pair of fins is far more beneficial than swimming for 30 minutes thrashing with no gear to assist.'

Training and distance

'The best advice I can give for social swimming is to progress slowly and set small goals. Swimming five kilometres on your first attempt is not going to happen. It is far easier to break things down into smaller sets and gradually increase the load as you gain confidence.

'For example, let's say you wanted to start swimming two kilometres every few days to train for a triathlon. The best way to start would be to break that 2000 metres into a smaller session, like 40 × 50 metres on 1.30-minute times. This gives you a chance to control your technique and breathing whilst still swimming a solid distance.

'Once you have managed the 50s with no issues, you can start mixing things up. There are infinite ways to change up a session, so have fun with it and challenge yourself. After six weeks the whole two kilometres will be a walk in the park.'

19

IS CARBON-FREE FLYING POSSIBLE?

Airlines promise to cut greenhouse gas
emissions, but how? Will air travel one day
be a 'green' adventure? What will fuel the
aircraft of the future?

Patrick Hatch

O ne of the most dramatic impacts of the COVID-19 pandemic happened in the skies: they fell silent. Airports sat empty and jetliners were mothballed in desert storage facilities protected from corrosion by the dry heat. By May 2020, global aviation had slowed to just 6 per cent of what it was before the pandemic. The best estimates are that it will take until 2024 for airlines to recover to 2019 levels of flying.

That makes COVID-19 the biggest setback to aviation since the Second World War, slowing what had been an inexorable growth in flying activity, which had tripled between 2000 and 2019. When the recovery does come, attention will again turn to one of the most pressing concerns about aviation: its impact on the environment and, specifically, its contribution to climate change.

While our homes, workplaces, cars and major industries are going green thanks to the uptake of clean electricity, there is no quick or easy replacement for the CO_2-spewing jet fuel that propelled 4.5 billion passengers through the skies in 2019.

So just how bad for the environment is flying? What's being done about it? And is it possible to fly 'carbon neutral'?

What effect does flying have on the environment?

In a normal year, aviation contributes about 2 per cent of the world's carbon emissions, according to the International Civil Aviation Organisation (ICAO), which is the United Nations' aviation body. Before the COVID pandemic, ICAO estimated that aviation's carbon emissions could triple by 2050, based on the projected continued growth in air travel.

In Australia, domestic flights (which are all that are counted in the government's National Greenhouse Accounts) made up 1.6 per cent of our total emissions in 2017. When international flights are included, emissions increase to represent 3.8 per cent of Australia's total. While that's a lot, it trails our biggest polluters – electricity and heat production (32 per cent of the total), road transport (14 per cent), and methane produced by farm animals (8.7 per cent).

One person flying economy class return from Sydney to London via Singapore will produce 1.74 tonnes of CO_2, according to the ICAO, which represents 8 per cent of our per capita emissions per year (21.4 tonnes per person).

Your carbon footprint doubles if you splurge for a business class seat because you take up more space on board, meaning the plane burns more fuel per person compared with a plane that has only economy seats (that's one reason to feel good being squeezed in the back cabin).

In the face of this, the world's major airlines (which are represented by a body called the International Air Transport Authority, or IATA) have pledged to stop growing their net carbon emissions from 2021, and by 2050 to have cut 'net' emissions to half what they were in 2005. Several airlines, including Qantas, have taken the bolder step of pledging to reach net zero emissions by 2050. Airlines face a huge challenge reaching these goals but insist they have the means to do it.

Several airlines, including Qantas,
have taken the bolder step of pledging to
reach net zero emissions by 2050.

How does carbon offsetting work?

For many people, the only time they have considered flying's contribution to climate change is when asked whether they want to pay extra to 'fly carbon neutral' as they book their ticket. If they say yes, they pay a few extra dollars on a Melbourne–Sydney flight or up to $50 on a return trip to London, and this money goes into environmental projects intended to mitigate the carbon impact of the journey. Not many people do this. Qantas says passengers paid to offset 133,242 tonnes of carbon through its schemes in 2018, amounting to only about 1.1 per cent of the airline's net emissions.

Climate scientists question the merit of some of the projects that supposedly undo the damage of your flight, many of which involve planting trees or conserving forests. While trees may suck carbon out of the atmosphere, they are part of the 'active' carbon cycle – they absorb carbon when they grow, which returns to the atmosphere if they fall down and decay, or are burnt. That doesn't mitigate the impact of releasing carbon that's been locked up in the ground for millions of years and is then burnt by airlines as jet fuel, these critics say.

The gold standard for offsetting are schemes that fund clean energy projects such as building wind or solar farms, which create electricity that would otherwise be produced with fossil fuels, and in that way genuinely offset the carbon from a flight.

Airlines will start relying more and more on offsetting schemes to mitigate their emissions, so that even though the amount of CO_2 their planes spit into the atmosphere will keep rising, their net emissions – that is, their actual emissions minus what they offset – will stay at 2019 levels and then eventually fall. For international air travel this offsetting is governed by a United Nations scheme called the Carbon Offsetting and Reduction Scheme for International Aviation (CORSIA), which will ensure each carrier has offset their carbon emissions growth properly. Australia is one of 80 countries, covering three-quarters of the world's international air travel, to take part in the CORSIA scheme from 2021 before it becomes mandatory for 191 countries in 2027.

Could biofuels become the new aviation fuel?

Rather than offsetting CO_2, it is, of course, better not to release it into the atmosphere in the first place. Airlines boast that they have already made great strides on this front, with per-passenger emissions halving since 1990 as carriers swap fuel-guzzling, four-jet planes such as the Boeing 747 jumbo with newer, more fuel-efficient aircraft such as 787 Dreamliners and Airbus A350s (the main driver being to cut their fuel bills rather than to slow global warming). They say planes will continue to get lighter and be able to fly further with less fuel.

Rather than offsetting carbon dioxide,
it is, of course, better not to release it
into the atmosphere in the first place.

The biggest opportunity to cut emissions in the medium term is to switch from jet fuel refined from petroleum to biofuels. These alternative fuels can be made from vegetable oils derived from seed crops, used cooking oil, algae and even household or forestry waste.

Biofuels can produce up to 80 per cent fewer emissions than conventional jet fuel without significant changes to aircraft engines or airport infrastructure. The industry specifies it is interested only in sustainable aviation fuels (SAFs) – which are biofuels made from crops that do not require land clearing, do not replace food crops and use water responsibly – to ensure the switch to green fuel does not cause yet more environmental harm. But fuel is already the single biggest expense for airlines and biofuels currently cost up to four times as much. So uptake has been negligible, and sustainable aviation fuels accounted for one out of every 10,000 gallons of fuel the global aviation industry guzzled prior to COVID.

Several airlines – including Qantas, Lufthansa, Cathay Pacific and United – have made long-term purchasing commitments to encourage biofuel production, while Norway has mandated that its airlines must use 30 per cent sustainable aviation fuel by 2030. Qantas in 2019 committed to spending $50 million over the next ten years to kick-start a local biofuel industry.

The International Air Transport Authority believes that biofuel uptake needs to hit 2 per cent by 2025 to reach a critical mass of production, after which the cost would start to fall. But it argues that won't happen without significant government investment to help things along. (Unsurprisingly, while calling for government help, airlines firmly oppose measures such as carbon taxes or levies that would financially incentivise them to speed up the transition to sustainable fuel themselves.)

Modelling commissioned by the British group Sustainable Aviation estimates production of sustainable aviation fuels could lift to between 14.5 and 30.9 million tonnes annually by 2035 – enough to replace only 4 to 8 per cent of the global industry's current jet kerosene use.

Airlines can use only a 50/50 mix of biofuels with conventional jet fuel because aviation authorities are not yet satisfied that engines can run as reliably and safely with a higher blend. But work is under way to change this. Airbus in March 2021 operated its first test flight using 100 per cent biofuel with an A350 jet, while Boeing announced in January 2021 that all its aircraft would be certified to fly entirely on biofuel by 2030.

Will planes one day run on hydrogen?

The aviation industry is looking beyond oil and even biofuel to a new fuel source that could eradicate carbon emissions from aviation entirely: hydrogen. The lightest element on the periodic table is increasingly seen as playing a vital role in addressing the climate crisis across a range of industries that cannot easily switch to wind or solar power.

Hydrogen is produced through a process called electrolysis

where electricity splits water (H_2O) into just the O (oxygen) and the H (hydrogen). When burnt in a combustion engine in liquid or gas form, hydrogen's only by-product is water vapour. Most hydrogen production today is powered with fossil fuels but renewable energy powers the green hydrogen that airlines would use if they were to fly carbon-free. It is estimated that more than $194 billion of green hydrogen projects were announced worldwide in 2020.

Hydrogen-powered flight is nothing new. It's the main fuel source used in space flight. In 1988, the Russian aircraft maker Tupolev experimented with liquid hydrogen in a modified version of its Tu-154 workhorse, used by Soviet airlines for decades. (They called the hydrogen option 'ecologically pure', although back then the reported aim was also to find alternatives to petroleum for when global supplies dwindled.) And Airbus led a European Union–backed study, in the early 2000s, into the viability of 'cryoplane' – a reference to the need to cryogenically cool liquid hydrogen to transport it. That study found that hydrogen could, with special precautions, be used just as safely as kerosene. But that did not translate into meaningful action.

The concept of hydrogen-powered passenger aircraft was given a major boost in September 2020 when Airbus released concepts for three zero-emission passenger aircraft that it says airlines could be flying by 2035. The planes are all intended to be fuelled by either liquid or gas hydrogen. One is powered with electric turbo-propellers, seating up to 100 passengers; another is a turbofan jet carrying up to 200 passengers; and the third is a jet-powered 200-seater with a revolutionary blended-wing body, which looks more like a kite or fighter jet than a conventional passenger aircraft. Airbus says it will make a decision by 2025 on which of the concepts to put into production.

Airbus hydrogen concept planes

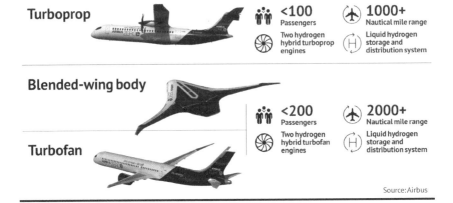

Turboprop

<100 Passengers

Two hydrogen hybrid turboprop engines

1000+ Nautical mile range

Liquid hydrogen storage and distribution system

Blended-wing body

Turbofan

<200 Passengers

Two hydrogen hybrid turbofan engines

2000+ Nautical mile range

Liquid hydrogen storage and distribution system

Source: Airbus

The hydrogen-combustion engines on the 200-seat jet and blended-wing models would eliminate CO2 from flying (provided they were powered by green hydrogen – that's the one made with renewables). The smaller propeller aircraft would be powered with hydrogen fuel cells – which work like batteries by converting hydrogen and oxygen into electricity – making it even greener because that would eliminate the greenhouses gases nitrous oxide and sulfur oxide.

The final piece of the puzzle is to reduce or eliminate 'contrails' – the white streaks of water vapour cloud that form behind a plane and affect the Earth's temperature by blocking both sunlight and radiation reflected back from the planet's surface. It is thought hydrogen engines will produce thinner and less persistent contrails because they will not emit soot particles for the water vapour to cling on to.

Despite the hype, a shift to hydrogen flying is far from straightforward. One of the biggest challenges is that although hydrogen fuel is very light, it has a very poor energy density by volume. That means a lot more space is needed on a plane to carry hydrogen fuel compared to standard jet fuel to fly the same distance. Special storage tanks will be required to cool the fuel to minus-260 degrees.

As a result, Airbus's three hydrogen concept planes are expected to have a shorter flight range than their petroleum-run equivalents. The manufacturer's 180-seat workhorse A320s have a range up to 6300 kilometres, and have been used on eight-hours flights from Sydney to Manila, for instance, but the turbofan hydrogen equivalent is set to be limited to about 3700 kilometres, which wouldn't get you much further than Darwin. Passenger jets today carry most of their fuel in their wings but concepts for new hydrogen aircraft have leant towards incorporating them in the plane's fuselage (the long tube part). This is where Airbus's new blended-wing aircraft concept comes into its own: it will potentially be able to hold enough fuel in its large frame to fly longer distances.

Another issue is getting hydrogen fuelling infrastructure in place at airports, so planes can be guaranteed they can refuel – a massive undertaking.

But don't try booking a seat on one of these whiz-bang hydrogen planes yet. It is unlikely that hydrogen aircraft will replace existing jetliners in any meaningful number until well beyond the middle of the century, given airlines generally hold on to aircraft for 20 to 30 years.

In the nearer term, hydrogen might play a role through the production of synthetic electrofuels. Created by mixing green hydrogen with CO_2 captured from industrial factories, coal or gas-fired power plants or pulled directly from the atmosphere, electrofuel could supplement other biofuels in order to lower emissions.

Meanwhile, airlines including Air New Zealand have already been looking at ways to electrify their propeller aircraft using conventional batteries for short-haul flights, with Norway planning for all domestic flights to jettison sustainable fuel and be electric by 2040.

And a group of aircraft manufacturers rushing to build the next generation of supersonic passenger jets, two decades after the end of the Concord era, have a low-carbon future in mind. British aerospace engineering firm Reaction Engines says its hydrogen-powered jets could propel aircraft from Sydney to Brussels in less than five hours, while Boom Supersonic plans for its 55-passenger jet – which it claims could be operational in 2030 – to be able to run entirely on sustainable biofuel.

So, should you avoid flying for now?

In Sweden, there's a word for feeling guilty about jumping on a plane – flygskam – or 'flight shame'. With no easy or quick solutions to aviation's climate problem, a growing number of people are choosing not to fly at all.

A European Investment Bank survey in January 2020 found that one in three Europeans said they already flew less for holidays because they were concerned about climate change, and three-quarters said they intended to fly less for that reason in 2020 : (not, as it turned out, that they had many opportunities to do so). So it's not surprising that other modes of transport are back in vogue. Travel on Sweden's largest rail network jumped 11 per cent in 2019 and rail operators across Europe are reviving overnight sleeper train services between major cities.

With no easy or quick solutions to aviation's climate problem, a growing number of people are choosing not to fly at all.

In a sign that airlines are conscious of this shifting landscape, Lufthansa, Air France and Dutch carrier KLM have partnered with rail operators to sell code-share tickets across both sky and land.

KLM launched a 'fly responsibly' campaign in 2019 encouraging passengers to save time and the environment by taking the rail link between Amsterdam and Brussels rather than flying, while Lufthansa axed its short-haul Frankfurt–Cologne flights due to the success of its alliance with Deutsche Bahn.

French lawmakers moved to accelerate this trend by voting in 2021 to ban all domestic flights on routes that can be covered by train in less than two-and-a-half hours, unless passengers are connecting to an international flight. So, no more flying from Paris to cities such as Nantes, Lyon and Bordeaux.

That followed Austrian Airlines ditching its 45-minute air service between Vienna and Salzburg in favour of more frequent train services with its rail code-share partner in mid-2020, after the Austrian government made cutting short-haul routes a condition of a €600 million ($940 million) pandemic bailout.

Of course, Australians don't have much choice if they want to travel abroad or even around the country. The absence of fast rail connections between our vastly separated cities makes us almost uniquely reliant upon air travel. The country has flirted with the idea of high-speed rail since the 1980s but nothing has ever advanced past the planning stage. A bullet train connecting Melbourne, Sydney and Brisbane would take up to 50 years to build and critics say the population of those cities is too small to support such a project (unlike the far larger, and closer, cities connected by high-speed rail in Japan, China and Europe). In any case, the airline body IATA warns that about 80 per cent of the industry's CO_2 emissions come from flights of more than 1500 kilometres, for which rail is not a viable replacement.

The long-term impact of COVID-19 on air travel is still unclear. Holidaymakers might be more inclined to travel closer to home in the future, and the post-COVID uptake of videoconferencing might eliminate some demand from business travellers. But whatever our travel plans, it's unlikely the climate clouds hanging over aviation will part any time soon.

20

WHAT IS IT LIKE TO LIVE WITH DEMENTIA?

Every three seconds, someone in the
world develops a form of dementia.
So why is it so poorly understood?

Jewel Topsfield

N atalie Ive was in her forties when she started forgetting the names of children in her kindergarten class. She loved being a teacher and quickly developed workarounds; checking the roll or a child's name on a painting to prompt her memory.

But her family was noticing changes, too. She had to consult recipes to make meals she once cooked effortlessly. Sometimes her children helped her cross the road because she couldn't work out how far away the cars were.

One day her daughters went to the movies. 'I just lost where I was,' Ive says. 'I had put eggs on to boil and the water all evaporated. The girls tried to call me, I didn't know how to use the mobile to answer it. I didn't know how to open a door or anything. I was just sitting there.'

Geoff Fairhall was a polymath. An academic and senior executive, he learnt seven languages, was passionate about music, the arts, architecture and design, flew light aircraft and was a keen sailor. And then, out of the blue, he began struggling to tie his boat to the trailer. 'It was confusing, I didn't know what was going on,' says his wife, Anne.

When Mithrani De Abrew Mahadeva was diagnosed in 2015, she was told to prepare for a nursing home. She was 65. 'I said, "I don't want to go into a nursing home, this is my own home, something I am familiar with."'

Her friends didn't believe the diagnosis. 'That's because I was talking,' she says. 'But I was struggling, I was stuck inside the house because I was having falls and memory problems.'

Ive, Fairhall and De Abrew Mahadeva all live with dementia, the second leading cause of death in Australia after heart disease, and the leading cause of death among women.

In 2021 there are 472,000 people in Australia living with dementia – about 2 per cent of the population – and someone in the world develops a form of it every three seconds. Almost one in ten Australians over 65 has dementia, rising to three in ten over 85.

Yet it remains poorly understood. Many people wrongly believe dementia is a normal part of growing old. And memory loss is not the only symptom: dementia can affect behaviour, mood, language, planning, problem solving, spatial awareness, sensory perception and the ability to experience pleasure.

So what exactly is dementia and what is it like to live with it?

What is dementia?

Dementia is not one specific disease but the umbrella term used to describe the symptoms caused by a group of brain diseases that affect both young people (younger onset dementia) and those over 65. It occurs when neurons (nerve cells) in the brain stop working and disconnect from other neurons. Everyone loses some of these connections between neurons as they grow older; those with dementia lose more.

> *Dementia is not one specific disease but the umbrella term used to describe the symptoms caused by a group of brain diseases.*

'Two of the biggest myths are that dementia is a normal part of ageing and only old people get dementia,' says Dementia Australia CEO Maree McCabe.

There are more than 200 different diseases that cause dementia, although most are very rare. The most common is Alzheimer's disease, which accounts for about two-thirds of cases. Other common forms include vascular dementia, Lewy body disease and frontotemporal dementia. Another disease that can cause dementia is chronic traumatic encephalopathy (CTE), which is linked to repeated blows to the head and was found post-mortem in former VFL/AFL players Polly Farmer, Danny Frawley and Shane Tuck.

Many people wrongly believe dementia is a normal part of growing old.

Of the almost half a million Australians who have dementia, 28,300 (about 6 per cent) are younger than 65. 'It's much rarer for younger people to get dementia, but it's certainly not as rare as we'd like it to be,' McCabe says. 'Often some of the younger forms of dementia that occur are inherited.'

Dementia is progressive and ultimately terminal, although a person can go on living for many years. In late-stage dementia, individuals may lose the ability to walk and speak. Swallowing difficulties increase susceptibility to aspiration pneumonia, caused by accidentally inhaling food or fluids into the lungs, which leads to many deaths. Others may die as a result of complications related to loss of brain function, dehydration, malnutrition, falls and infections.

What is it like to live with dementia?

People living with dementia often find that everyday situations – making cups of tea, remembering instructions and going to the toilet in the middle of the night – become more challenging.

Natalie Ive, now 49, was diagnosed with younger onset dementia/primary progressive aphasia, a type of dementia that makes it difficult to express thoughts or find words.

She meticulously war-games trips to the supermarket. She carries a shopping list, ticking off each item in the trolley, and listens to music through headphones 'to block out all the other stuff so it doesn't become overwhelming'.

'[Otherwise] there are just too many sensory things going on,' she says. 'I've got my notepad, I've got my pen, music going, so I am just focusing on that. Tick, tick, tick.'

As a teacher, Ive excelled at communicating. Now it's sometimes hard to get her thoughts across. 'But I know that inside there is that determination. This is who I am, I know I am able to do it, but maybe not in the same way.'

She uses strategies to help her understand and express herself:

requesting frequent breaks, asking for information to be repeated and broken down into single steps, taking big breaths and pausing. 'That allows for me to be able to continue to the best of my capabilities.'

Ive stresses that she continues to live a fulfilling life. She enjoys painting, diamond art (a form of mosaic), gardening and beach walks. 'The beach gives me great calm, it centres me. It's very therapeutic, when I go there it just helps with some clarity.' Instead of reading, she now listens to audio books.

Her mission is to educate the community about dementia. She would love to see dementia-friendly hours at supermarkets (similar to the Quiet Hour at some Coles stores, where noise and distractions are reduced) and has inspired her gym to fundraise for dementia.

'I want to really highlight the positives,' she says. 'Even on bad days I am still doing things for myself. Yes, aphasia is a part of me, but it doesn't define me.'

De Abrew Mahadeva was diagnosed with mixed dementia (Lewy body disease and Alzheimer's) when she was 65. Now 71 and living in her home, where she is cared for by a daughter, she has 'my very good days' and then 'really bad' nights, when she is kept awake by traumatic memories.

Her handwriting is getting messier, she says, and sometimes when she reads in the morning she knows she is looking at letters but can't make out the words.

To give others a sense of what it's like to live with dementia, Dementia Australia uses virtual reality. Strap on goggles and you see the world through the eyes of Edie, a man living with dementia who has to go to the toilet in the middle of the night.

It's wildly disorienting: the walls rustle like a living creature, furniture rushes at you at distorted angles and the toilet – a muted blur of shapes in the dark – turns out, too late, to be the washing basket.

'We'll have family members say, "Look, I don't get it, Mum or Dad walks into the bathroom where the toilet is and they wee in the corner, why don't they wee in the toilet?"' says McCabe. 'Well, they can't see the toilet. So the toilet seat needs to be a different colour so that they can actually see it.'

Another virtual-reality scenario looks at mealtimes. 'Sometimes you'll have a white plate on a white tablecloth and they might have chicken and cauliflower and peas, and they'll only eat the peas, because that's all they can see,' McCabe says. 'So having coloured crockery shows up the food on the plate and enables them to see what they're eating and choose what it is that they want off their plate.'

What is it like living with someone who has dementia?

The diagnosis often comes as a relief, although there is also a grieving process.

Chris White was a high-profile union leader for 40 years. As secretary of the Trades and Labor Council in South Australia, he argued the state wage case, pushing for an increase in the minimum wage. He kept detailed diaries and was a persuasive public speaker, featuring regularly in the media.

'Now he can't organise appointments and he also struggles with recording them,' says his partner, Kathryn Moyle. 'He doesn't even remember emergency numbers. So he went through an enormous grieving process, as did I, and at one stage he said, "I'm no longer Chris, you can call me somebody else because I'm not Chris any more."'

Moyle noticed changes a decade before White was diagnosed five years ago with frontotemporal dementia at the age of 67. 'I had been saying for years to a friend, "I really don't think he is the same, there is something wrong."'

White began sleeping a lot in the afternoon and behaving in an antisocial way that was out of character. Once he loudly told two people they were too fat to fit in a lift.

'The diagnosis was really comforting at one level and distressing at another,' Moyle says. 'I was relieved to know that this is what was wrong with him, because he didn't accept I thought his behaviour was problematic.'

People with the behavioural variant of frontotemporal dementia often have trouble controlling their behaviour and may say inappropriate things or ignore others' feelings. However, they rarely notice these changes. 'One of the things that I spend a lot of time doing is apologising to people,' Moyle says.

'For example, if Chris doesn't get what he considers to be the right answer at the library when he wants to borrow a book, he will just get really angry and start yelling, and the behaviour is completely over the top for the issue. So he actually needs to be accompanied most of the time and that generates its own challenge for me.'

She's had to become 'extraordinarily patient'. It takes three attempts to get out the door. 'Wallet and hearing aids are often misplaced,' Moyle says. 'There's a real line between hyper-attentiveness that is so controlling it stifles and trying to position Chris to keep his independence.'

Moyle misses the confidential conversations and philosophical debates you have with your partner. 'Now Chris gets agitated easily, those sorts of conversations are no longer there for me. That being said, I am glad I retired in 2019, because it means that we have a lot of time together. And when Chris is firing on three out of six cylinders – which is the best we can expect – it's lovely to have him around, he's charming, he's voracious, he wants to go to the theatre, he wants to go to the galleries.'

Geoff and Anne Fairhall went on a round-the-world trip in 1991. Geoff, then 51, struggled to cope with things that normally wouldn't faze him, such as delayed flights and hotel alarms going off in the middle of the night.

'He was really cranky in London,' Anne Fairhall says. 'I put that down to being tired but it was unusual because he was a gentleman and always very diplomatic and controlled in a dignified and calm way. This was a public display of anger.'

She says he became 'quite disinhibited in all sorts of difficult ways'. Once, he hit her on a plane. She was flummoxed because it was so utterly out of character for her gentle husband. 'He had never whacked me in my life.'

Geoff was finally diagnosed with frontotemporal dementia at the age of 66, after years of incorrect diagnoses, including anxiety and depression. One doctor even told him he was 'too young' and 'not the type' to get dementia. 'It was terrible – in those days doctors didn't have a good understanding,' Anne says.

Family carers often experience a particular kind of grief. 'It's not like an end-of-life grief, where it's very final, it's a living grief and it's an anticipatory grief,' she says.

'Some people say "Well, he hasn't died", but you can still go through the phases of grief even when the person hasn't died, because you've lost a person you did have. But you need to concentrate on the things you can do, not the things you can't do.'

How do carers cope?
Anne Fairhall recalls creeping into the offices of Dementia Australia and being surprised by how normal everything looked.

'Maybe I thought it was going to be like a leprosy hospital or something – I thought I had been tarred with something that was shameful, because there is a lot of embarrassment.'

The organisation offers support for families and carers and she was impressed with how helpful it was. She attended a course for couples who had just received a diagnosis, where she met other carers and learnt about services available.

During the middle phase of Geoff's dementia, she says the stress was 'unbearable at times': 'disturbed sleep, collapsing in the bathroom, going out the front door at 1am and not knowing when they're going to come back.'

Aggression can be an issue with frontotemporal dementia and Fairhall had a bag permanently packed so she could leave the apartment in times of emergency. She coped by reminding herself this was the disease and not her husband. 'This is not who he is, this is an aberration. I've known him long enough to know he loves me, I love him.'

Moyle recommends carers maintain their own interests and friendships and ask for help. 'There is a danger of your being absorbed by it, it can be totally and utterly overwhelming,' she says. 'I do think it's really important for carers to have social workers and counsellors who are on your side.'

She advises telling people about the diagnosis because dementia is so complicated and misunderstood. 'When Chris was diagnosed we decided we would just tell a few people and then we decided that was ridiculous because it was asking people to keep a secret. We also felt it would be better to keep our friends in the loop if Chris started behaving weirdly.'

Do people treat you differently if you live with dementia?

There is a scene in *The Father*, which earned Anthony Hopkins an Oscar for best actor, in which a carer tries to coax him – a man living with dementia – into swallowing a 'pretty blue' pill. The man snaps at her, telling her not to talk to him as if he's 'retarded'.

Natalie Ive's daughter, who was watching the film with her, dug her in the ribs. 'She's like, "That's you!"' Ive says.

Ive rails against being treated differently and finds it irritating when people think of her as suffering. 'I'm not suffering, it's quite the opposite. I'm struggling, in a sense, but I am not suffering. I'm Nat, I'm still Nat. Okay, there are difficulties and challenges, but I can do things in a different way.'

A survey released by Dementia Australia in 2020 found discrimination was a real issue for people living with dementia. The survey of more than 5700 people found three-quarters of those who live with dementia say people don't keep in touch like they used to. Sixty-five per cent said people they knew had been avoiding or excluding them. Ninety per cent of carers, friends and family members felt people with dementia were treated with less respect and 81 per cent felt they were treated differently in shops and restaurants.

McCabe says discrimination stems from a lack of understanding and knowledge of dementia, what it is and how it impacts people.

'A little bit of support can really make a really big difference. It could be as simple as giving someone space to do things for themselves, listening to the person, not trying to solve all their problems, giving the person time to find the right words or using technology to support someone in their day-to-day activities.'

What shouldn't you say to someone living with dementia?

'Remember?' or 'I already told you.'

'When you say to somebody "Don't you remember? We've had this conversation 30 times", what it highlights for people is their deficit. It's absolutely confronting for them,' says McCabe.

'It's better to just give them the answer to the question. They're not doing this to drive people crazy, they're doing this because it is the first time they're asking, in their world.'

Some people living with dementia rely on confabulation – the subconscious creation of false stories – to fill gaps in their memories. They are unaware that what they are saying is untrue.

Anne Fairhall was recently greeted by staff at Geoff's aged care home who said they had heard all about how the couple met in Nepal. 'I said, "Oh, that would have been nice."' Fairhall says. 'We met on the beach at Ocean Grove, nowhere near as exotic. You've got to have a sense of humour, because sometimes some of these situations are really, really funny to our normal brain but you have to respect their dignity and the fact it's very real to them.'

It's also advised not to argue with someone living with dementia because it could upset them or make them angry. 'You're constantly swallowing your pride because you don't get to ventilate what your view of it is,' says Kathryn Moyle. 'But there's no value in having that kind of discussion . . . Chris can't cope with any level of conflict any more.'

It's also not uncommon for someone living with dementia to believe that a loved one who has died is still alive. Should you tell them the truth?

> *You have to respect their dignity*
> *and the fact it's very real to them.*

'As humans we are prone to reinforce the truth, inherently we want to be honest,' McCabe says. 'But it's not always the best thing for people living with dementia. Every time you tell them, it's like the first time they hear it, it's cutting, it's devastating. So often it's best to say "Look, they're just not here today" or "We'll catch up with them later."'

So what should you do?

Looking at photos or articles can help spark memories. Geoff Fairhall grew up in Tasmania and the couple honeymooned on Lord Howe Island, so Anne will often bring in mementos from these places when she visits the aged care home.

'Geoff will remember things that go back 40 to 50 years ago, whereas if you asked him about something that happened five minutes ago, he's lost it,' she says. 'Photos can remind people of something earlier in their life without you having to say, "Remember when."'

She will always bring chocolates, which Geoff loves, and the couple will read together. 'He still has the ability to read in the moment,' she says.

Dementia Australia encourages grandchildren to visit, too. The organisation suggests playing games, watching a well-loved video, listening to music, helping with personal grooming and outings such as a stroll or short drive.

Can you prevent dementia?

No – but there are ways to reduce the risk.

A landmark report in medical journal *The Lancet* in 2020 identified 12 risk factors that, if addressed, might prevent or delay up to 40 per cent of dementias.

These risk factors include less education in early life, hearing loss, smoking, high blood pressure, obesity, depression, physical inactivity, diabetes, infrequent social contact, excessive alcohol consumption, head injury and air pollution. So increasing physical activity, maintaining a healthy, balanced diet, quitting smoking, limiting alcohol intake, improving sleeping patterns and checking your hearing can all help.

'Many people don't know hearing loss can increase your risk of developing dementia,' says McCabe.

Studies have shown that the brains of people with hearing impairment atrophy more quickly. The brain must also work overtime to understand what people are saying, so there are fewer resources available for other functions such as learning and memory. And people with hearing impairment are more likely to become socially isolated, intensifying the risk of dementia.

The *Lancet* report found untreated hearing loss was the largest modifiable risk factor, accounting for 9 per cent of cases. It recommended the use of hearing aids to reduce the risk.

What treatments are available?

A number of drugs are available in Australia for use by people with dementia. These include drugs that are used to improve memory, known as cholinergic treatments (Donepezil, sold under the trade name Aricept, is an example) and medication to treat some of the challenging behaviours.

'Cholinergic drugs are not very effective, they have lots of side-effects, but they are currently the standard of care for Alzheimer's,' says Associate Professor Michael Woodward, director of Austin Health's Memory Clinic and honorary medical adviser to Dementia Australia.

He says a low dose of an antipsychotic can be useful in extreme cases, such as when someone is taking 'swipes at hallucinations'. 'We tend to try and use those drugs very, very sparingly; they have certainly been overused, particularly in residential care.'

The first new treatment for Alzheimer's disease in 20 years was approved for use by the US Food and Drug Administration in 2021. The drug, known as aducanumab, removes amyloid plaques that accumulate in the brains of people with Alzheimer's and slows the progression of the disease. It has been trialled in Australia and throughout the world.

Woodward says it is very significant that aducanumab met the FDA's standards for accelerated approval, which is used to fast-track promising new drugs to treat serious or life-threatening illnesses. 'They felt there was enough evidence from the data that was presented to them to believe that this drug might be clinically beneficial,' he says. However, he says it still needs to be approved by the Therapeutic Goods Administration before it can be used in Australia.

'There's several other drugs that remove amyloid that look equally promising that may well end up also being approved. But they'll probably only work significantly if they're used early enough ... not in the advanced stage of dementia.'

Still, Woodward is optimistic he will see a cure for dementia in his lifetime. 'I've been saying it's five years away for the last 20 years,' he says.

Meanwhile, people living with dementia and family carers say it is possible to continue living well with dementia.

In 2020, Natalie Ive was one of 26 women who had achieved success in their field selected to walk the Priceline Beauty runway at Melbourne Fashion Week. 'I just felt so proud, so honoured, to represent younger onset dementia in front of the TV with other inspirational women. You don't need words with that. I'm able to do things and have the world watch.'

It's now 30 years since Geoff Whitehall started showing signs of dementia on that round-the-world trip. For some years after his diagnosis he was able to continue to do short-term projects in academia, sail, go on overseas trips and attend music concerts. His wife Anne says an early diagnosis is critical because it helps people with dementia and their families adjust and manage their lives.

When Geoff began to struggle to follow classical music, for example, he switched to jazz, which is more improvisational, and which he still finds calming. When travel became too challenging, Geoff and Anne's children, who were living overseas, visited them in Melbourne instead.

'You live in the present a bit more and look for moments of creating joy,' Anne says. 'People have always seen dementia as a death sentence, but you can go on living well with dementia for many, many years.'

*

For more information, go to dementia.org.au

21

HOW HAS SHIPPING CHANGED OUR LIVES?

Container ships navigate choke points and
even dodge pirates to bring goods to our
homes. But business is changing for the
unsung heavy lifters of global trade. How?

Matt Wade

When the 400-metre long, 220,000-ton container ship *Ever Given* got stuck in the Suez Canal in March 2021, global shipping was thrust into the limelight. We don't normally pay much attention to international sea lanes but the Suez snarl, which lasted for six days, underscored how dependent we are on the maritime arteries of international trade.

By blocking the busy shipping lane that connects Asia and Europe, the prone vessel halted the passage of almost US$10 billion worth of seaborne traffic per day – around US$400 million each hour. The disruption to trade was still being felt weeks after the hulking *Ever Given* was finally dislodged.

At any given moment, tens of thousands of commercial vessels are plying the world's oceans, the unsung heavy lifters of the global economy. 'We associate airports with air travel, which is all very glamorous and linked with holidays and so on, whilst shipping does the day-to-day grunt work of global trade,' says Tim Harcourt, a trade economist at the University of Technology, Sydney. 'Shipping is like the hard-working midfielder and airlines are like the fancy full-forwards.'

Even before the *Ever Given*'s mishap, disruptions caused by the pandemic meant international shipping was under pressure. The flurry of online purchases we made during lockdowns triggered a surge in demand for the giant vessels that traverse the high seas laden with cargo. In April 2021, the price of shipping a container from Asia to Australia was about 150 per cent higher than before the pandemic, says Marika Calfas, chief executive of NSW Ports, which runs Sydney's Port Botany.

Despite the scale and sophistication of global shipping, modern seafarers still face some age-old hazards – there were 195 incidents of piracy against ships worldwide in 2020, 20 per cent more than in 2019. And some crucial seaways used by the behemoths of maritime trade are increasingly contested by regional and global powers.

So how did we come to rely so heavily on maritime trade? How vulnerable are the sea routes that keep the global economy going? And how is global shipping changing?

Who invented container ships?

A simple idea revolutionised the shipping industry in the mid-1950s. American truck driver Malcolm McLean stacked 58 metal boxes on an ageing tanker ship going from Newark on the US east coast to Houston, Texas. This concept sparked a flurry of innovation, including a standardised, truck-sized container called 'twenty-foot equivalent units' or TEUs. Shiploads are measured in TEUs, but containers now come in several sizes: three metres and six, twelve and 13.7 (in feet, that's 10, 20, 40 and 45). Given their size – designed to fit on trucks – many unused shipping containers have been recycled into small houses, granny flats and sheds.

Prior to McLean's invention, most shipped items were packed in barrels, sacks, baskets, crates or pallets then loaded and unloaded separately, partly on the backs of wharfies. It was a slow, labour-intensive and backbreaking business. But the introduction of the shipping container brought sweeping changes to international trade by slashing transportation costs.

In his 2006 book *The Box,* economist Marc Levinson explains how the standard-size container allowed huge economies of scale because ships, port facilities, trucks and trains in every country could be purpose-built to take any container in the world. The lower cost of shipping meant more factories could be located a long distance from customers, paving the way for economies in Asia, especially China, to become global manufacturing hubs. The introduction of refrigerated containers allowed perishables such as fruit, vegetables, meat, dairy, flowers and some pharmaceuticals to be transported to distant markets.

As a result, containerisation has been an important factor in the advance of globalisation. Cheaper, more efficient shipping has underpinned the development of sophisticated global supply chains and the 'just-in-time' management strategies embraced by manufacturers, retailers and others. Rather than incur the costs of stockpiling goods in warehouses, companies rely on the global shipping industry to deliver what they need when they need it. This has given households access to a vast array of low-cost products – everything from power tools to iPhones and fresh fruit to plastic toys.

Around 90 per cent of the world's traded goods are transported by sea on a variety of ships. Tankers carry liquid cargo, mostly oil, while dry bulk carriers move huge quantities of commodities such as grains, coal and ore. Much of those raw materials are taken to manufacturing regions where they are made into finished goods, which are themselves then moved back across the oceans in container ships or more specialised cargo vessels such as the 'roll on roll off' transporters that carry vehicles. The UN's Conference on Trade and Development (UNCTAD) says the total value of the world's merchandise exports reached US$19 trillion in 2019.

Around 90 per cent of the world's traded goods are transported by sea on a variety of ships.

A huge workforce keeps that trade moving – about 1.5 million seafarers are employed by the global shipping industry and each month about 150,000 crew members need to be changed over to, and from, the vessels they operate.

China is now at the centre of shipping commerce, especially container cargo. It hosts the world's biggest container port, in Shanghai, which moved 42 million containers in 2018. By comparison, all of Australia's container ports combined move around 8 million per year, mostly in Melbourne and Sydney.

The increasing scale of tankers carrying oil and gas from giant terminals and the bulk carriers transporting grains, coal, ore and cement has also helped to drive global trade growth. Tankers and bulk carriers are fundamental to the world's supply of food and fuel. These giant vessels share the high seas with other commercial vessels – fishing boats, passenger liners, ferries and so on – and an array of more specialised vessels that provide specific maritime services such as drilling, research, salvage and dredging.

How does shipping affect our daily lives?

The effect of shipping-based innovations on the everyday lives of Australians has been profound. More than 42 per cent of goods in a Sydney household arrive in containers through Port Botany, according to research commissioned by NSW Ports. Each year the Port of Melbourne, which is the biggest container port in Australasia, handles nearly 3 million standard containers.

Container shipping is now fundamental to our society, says NSW Ports's Calfas. 'It's integral at a personal level, a family level and at a business and economic level,' she says.

And yet commercial shipping and ports receive surprisingly little public attention. Michael Bell, professor of Ports and Maritime Logistics at the University of Sydney, says that reflects positively on their efficiency. 'From the point of view of the consumer, it works; the goods turn up and they are on the shop shelves,' he says.

Australia is especially dependent on international shipping. During the past three decades, our economy has become deeply integrated into global commerce, so much so that one in five of our workers is now involved in trade-related activities. We rank fifty-fifth in the world for population but a 2019 Department of Foreign Affairs and Trade report said Australia was the world's twenty-third-largest exporter and twenty-first-largest importer

The effect of shipping-based innovations on the everyday lives of Australians has been profound.

(although some of that trade is in services rather than the goods transported by sea and air).

Shipping allows Australia to earn income from agricultural and mining commodities that are far too plentiful for us to consume ourselves. Australia exports about two-thirds of its agricultural produce and most of its iron ore and metallurgical coal production (used for steelmaking). Most of Australia's merchandise exports leave on tankers and bulk carriers but when it comes to container shipping we import much more than we export. That means hundreds of thousands of empty containers are loaded onto ships each year and returned, mostly to Asian ports. 'Essentially, our biggest container export is air,' says Calfas.

> *Shipping allows Australia to earn income from agricultural and mining commodities that are far too plentiful for us to consume ourselves.*

The system can be perplexing for outsiders. Many commercial ships are registered under a flag that does not match the nationality of the owner. For example, at the beginning of 2020, more than half of all ships owned by Japanese entities were registered in Panama; more than a fifth of the ships owned by Greek entities were registered in Liberia and another fifth in Marshall Islands.

Bell says the main reason for this is that owners wish to avoid the stricter marine regulations imposed by their own countries, including labour rules, pay rates and taxation. Nations such as Panama and Liberia also offer simple and inexpensive registrations. But Bell says the quality of construction and maintenance of commercial ships is safeguarded by the need for insurance and the threat of inspection when vessels are docked at many foreign ports.

How secure is global shipping?

Problems created by the pandemic and the logjam in the Suez Canal caused by the *Ever Given*'s mishap show how vulnerable to disruption global shipping can be – and they have underscored concerns about the fragility of complex supply chains and just-in-time management strategies.

The Suez Canal, a narrow sea-level waterway built in Egypt during the nineteenth century, is a shortcut between the Mediterranean Sea and the Indian Ocean that means vessels travelling between Europe and Asia don't have to sail around Africa, saving weeks each journey. It is one of three strategically sensitive passages in the Middle East that carry a large volume of maritime traffic. The other two are the Strait of Hormuz, linking the Persian Gulf to the Gulf of Oman, and the Bab al-Mandab Strait, which separates Africa and the Middle East. All three are primary waterways for the transport of oil and natural gas.

The Panama Canal, opened in 1914, is a shortcut between the Atlantic and Pacific oceans. As with the Suez, it has been upgraded in the past decade to allow the passage of larger vessels. After the *Ever Given* debacle, Egypt announced it would further widen and deepen the Suez.

Another key maritime choke point is the Malacca Strait, a sea channel between Malaysia and Indonesia, which is the quickest route between the Indian and Pacific oceans. There is little margin for error in this congested shipping lane, which narrows to just 2.7 kilometres at one point. It is a natural bottleneck with potential for collisions, groundings and oil spills. If the Malacca Strait were blocked, almost half of the world's shipping fleet would be required to reroute around the Indonesian archipelago, adding to transport costs.

Strategic competition and diplomatic tensions are a perennial menace to commercial shipping, especially near these strategic choke points. One hotspot is the South China Sea, where China's growing military power and assertiveness has stoked international concern. Around one-third of global shipping travels through this sea, much of it via the Malacca Strait.

Choke points and hotspots

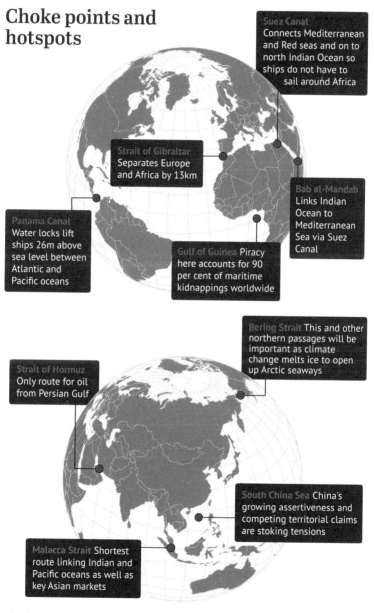

Suez Canal Connects Mediterranean and Red seas and on to north Indian Ocean so ships do not have to sail around Africa

Strait of Gibraltar Separates Europe and Africa by 13km

Bab al-Mandab Links Indian Ocean to Mediterranean Sea via Suez Canal

Panama Canal Water locks lift ships 26m above sea level between Atlantic and Pacific oceans

Gulf of Guinea Piracy here accounts for 90 per cent of maritime kidnappings worldwide

Bering Strait This and other northern passages will be important as climate change melts ice to open up Arctic seaways

Strait of Hormuz Only route for oil from Persian Gulf

South China Sea China's growing assertiveness and competing territorial claims are stoking tensions

Malacca Strait Shortest route linking Indian and Pacific oceans as well as key Asian markets

Illustration by Matthew Absalom-Wong

*Strategic competition and diplomatic tensions
are a perennial menace to commercial shipping,
especially near these strategic choke points.*

Piracy also poses a threat to commercial shipping in some regions. The industry is especially worried about deteriorating security in West Africa's Gulf of Guinea, which accounts for around 90 per cent of the world's maritime kidnappings. 'Previously, many of these attacks had been principally motivated by the intention to steal cargo,' says a report by peak body the International Chamber of Shipping. 'Increasingly, however, seafarers are now routinely being kidnapped and taken into Nigeria where they are then held for ransom in the most appalling and terrifying conditions. Most ship types have been targeted, including container ships and bulk carriers as well as tankers and offshore support vessels.' In the first quarter of 2021, 40 sailors were kidnapped globally, almost twice the number in the corresponding period the year before.

As with many industries, the coronavirus crisis has been hugely disruptive to seaborne trade. The seafarers who operate the global shipping industry were especially hard hit. Widespread border restrictions prevented many maritime workers from being repatriated during the pandemic and often stopped ship crews from being replaced. This forced hundreds of thousands of seafarers to continue serving on ships for months beyond their contracted tours of duty. The International Chamber of Shipping says many crews were pushed to near breaking point due to fatigue and the anxiety of being unable to return home.

How is global shipping changing?

When you stand on a dock alongside a modern container ship, its mammoth scale is striking. The distance from bow to stern of the world's biggest container vessels is more than double the widest point of the MCG's playing field. The economies of scale offered by even bigger vessels is appealing to shipping firms because the larger the ship, the lower the unit cost. Bigger vessels may also help shipping companies reduce their carbon footprints.

But Bell says further growth in ship sizes may be limited because it means more inventory must be stored at different stages of production in global supply chains and that costs money. 'Around 400 metres seems to be the maximum practical length,' he says.

Port operators complain that bigger ships require additional onshore investments to unload, store and transport their cargo.

Calfas says 'bigger is not always better' when it comes to container ships because the landside investments needed are likely to more than offset other savings. 'It's the shipping version of deploying an A380 to every airport,' she says.

Most of the container ships that come to Australia are smaller than those operating at the world's biggest ports. Calfas says the average container ship using Australian ports carries around 6000 containers, less than one-third of the capacity of the *Ever Given*.

Regardless of how big commercial ships become, advances in automation and other new technologies will affect demand for seaborne trade. Bell says digitalisation and artificial intelligence will allow far more localised production in some industries, which in turn will reduce the need to ship so many goods such long distances. 'What automation does is make production more mobile and less dependent on large skilled workforces,' he says. 'That means you can make things more easily closer to the point of consumption, and that's not good for shipping. New technologies give us a flexibility that means we don't have to source so many things from the other side of the world.'

A 2019 study by consultancy firm McKinsey found the intensity of goods trade is already declining in many global supply chains. Between 1990 and 2007, global trade volumes grew 2.1 times faster than gross domestic product on average – but between 2011 and

2019, they grew only 1.1 times faster than GDP. The study forecast advances in robotics, artificial intelligence and 3D printing were likely to 'reduce global goods trade by up to 10 per cent by 2030 as compared to the baseline'.

> *Regardless of how big commercial ships become, advances in automation and other new technologies will affect demand for seaborne trade.*

These technologies may lead to more 'nearshoring' – when businesses switch to suppliers in countries and regions close by in preference to those further away – allowing tighter coordination of supply chains and lower shipping times and costs.

Worries about the security of far-flung supply chains, stoked by the pandemic, could also encourage firms to reduce their use of long-distance shipping. The prospect of greater geostrategic tensions and trade conflict could be a further incentive for firms to reduce their dependence on distant suppliers and to source more goods closer to home – or from nations and regions considered friendly and less vulnerable to political influence.

As with most industries, shipping is under pressure to help combat climate change. The oil burnt to move so many huge cargo vessels around the world is a major source of the carbon emissions that cause global warming. International shipping accounted for around 2.5 per cent of global CO_2 emissions in 2019 – more than double what Australia produces.

The International Maritime Organization, the UN body that regulates international shipping, has set a target to reduce the carbon intensity of shipping by 40 per cent compared with 2008 levels by 2030 and by 70 per cent by 2050. Bell says some major players in global shipping are moving ahead of the regulator by experimenting with cleaner fuels and exploring new low-emission engine technologies. 'I'm quite optimistic about shipping making a transition to carbon neutrality,' he says.

While the industry might be trying to cut emissions, in one respect, at least, it might benefit from climate change. As the Arctic ice cap recedes, oceans once the domain of ice-breakers will host other types of ships. Waters in the region will remain navigable for much longer periods, creating potential new trade routes. Unfrozen, the North-West Passage, skirting Canada and Alaska, and the Northern Sea Route, above Russia and Scandinavia, could slash voyage times between Europe and key ports in Asia and North America. An experimental trip in February 2021 saw a Russian tanker become the first commercial vessel to cross the Arctic so far into in the northern winter. It provided a glimpse of the potential of Arctic sea routes as an alternative to the longer, traditional voyage between Asia and Northern Europe via the Indian Ocean and the Suez Canal.

While the industry might be trying to cut emissions, it stands to benefit from climate change.

But as nations jockey for influence over Arctic sea lanes, and for control of the vast natural resources below, greater geopolitical friction looms. There is also pressure to limit economic activity in the region in a bid to preserve the region's unique and sensitive environment. With climate change comes both concerns about more international disputes and the promise of a new frontier for maritime trade.

22

IS TIME TRAVEL POSSIBLE?

What is the twisty truth about spacetime? Will
we one day be able to shuttle into the future?
And what about visiting the past?

Sherryn Groch

US physicist Ron Mallett has wanted to turn back time since he was ten. That was the year he read H.G. Wells' novel *The Time Machine*. And the year he lost his young, clever father to a sudden heart attack. 'I decided I had to build a time machine so I could go back and see my father,' Mallett says.

In the opening pages of Wells' novel, he found a phrase that has never left him: 'Scientific people know very well that time is just a kind of space ... and we can move forward and backward in time just as we can move forward and backward in space.'

'It said *scientific* people,' Mallett says. 'We were plunged into poverty after Dad died ... and I went from being a really happy kid to really depressed – it just shattered me. But I knew I had to get to university, I had to become one of these scientific people if I was ever going to find a way to go back.'

Ron Mallett (standing) with his parents, Dorothy and Boyd, and brother Jason in 1948. After his father's sudden death, Mallett dedicated his life to the study of time and space. Courtesy Ron Mallett

Time travel might still sound like fiction, a problem for the likes of Marty McFly perhaps. But ever since Albert Einstein showed it was at least theoretically possible, in a series of equations that changed how we understand the universe, researchers have been giving the concept serious thought.

> *'Einstein died the same year as my father: 1955.*
> *They were the two giants of my life.'*

Now, after a long academic career studying black holes and lasers, Mallett thinks he has found a way to bend time. 'It came from solving some of Einstein's field equations. Einstein died the same year as my father: 1955. They were the two giants of my life.'

So what would time travel look like if it was cooked up in a lab rather than Hollywood? Is it logically possible to return to the past – or change it? And what does all this have to do with space travel and wormholes?

Theoretical physicist Ron Mallett with Einstein's field equation, which helps explain time travel. Courtesy Ron Mallett

What is time?

You are travelling in time right now but at the same boring old clip: second by second. Reality has a direction, things change from one moment to the next – ice melts, trees grow – and so we measure that passage by 'keeping time' with clocks.

Our experience of time is subjective. Sometimes it seems to speed up. At other times, perhaps even now while reading this, it feels punishingly slow, each moment dragging out longer than the next. But it's not always just in our heads – under the right conditions, time can change speed.

Ten years after Wells wrote *The Time Machine* in 1895, Einstein's seminal theory of relativity ruled that time and space really are part of the same fabric of the universe: spacetime. You can move in three dimensions in space (up/down, left/right and forwards/backwards) but there is a fourth dimension you need to locate yourself in, too: time. And as with space, time is malleable. In the words of one of television's most famous time travellers, Dr Who, this is where it all gets a bit 'wibbly-wobbly, timey-wimey'.

Time has a beginning, Mallett says: the birth of the universe in that first cosmic explosion known as the Big Bang billions of years ago. But scientists (and philosophers) are split over whether there is already an end – a future stretching out in front of us as real as the present and the past. Your death may already exist somewhere in spacetime, the theory goes. It imagines all of time and space as a block with no global or fixed present, where the difference between Cleopatra's reality back in ancient Egypt and yours now is just a matter of perspective.

From the halls of the University of Sydney's Centre for Time, philosopher Kristie Miller explains: 'It's like one big Persian rug, there might be dinosaurs up one end and sentient robots down the other, but it's all woven in already.'

And if the past and the future are both as real as the present, with plottable coordinates in spacetime, then why couldn't you change where you are in the block? If the past really is a foreign country, as the saying goes, couldn't we visit?

'The block theory of the universe makes time travel theoretically possible,' Miller says.

Of course, our intuition also rails against this idea. We know there is something inherently different about the past, the future and the present. We experience the present, it is what we see and feel and smell right now; we have memories of the past, ghost impressions of a former present, and we have none at all of the future. Doesn't that suggest it hasn't happened yet?

To account for this, some theorise the block of time is unfinished, that it grows from one moment to the next and has done so since the Big Bang. But the slippery science of trying to pin down any kind of universal present, that cliff-edge from which the bridge of reality is then built, renders this surprisingly inelegant. Scientists have largely given up on finding the cut-off point, Miller says.

But in looking for it, they found the first kind of time travel – forward.

Can we travel fast enough to reach the future?

To travel to the future, we typically imagine time speeding up. 'Night again, day again, faster and faster still,' Wells wrote, as the years whizzed by outside his fictional time machine. But to skip ahead, you actually want time to move more slowly for you, relative to everything else.

Einstein showed that two things can slow the usual flow of time: high speed and high gravity.

In the case of speed, the rule goes like this: the closer an object comes to travelling at the speed of light (299,792 kilometres per second), the slower time will pass for it. In 1971, scientists flew atomic clocks, the world's most accurate timekeepers, around the Earth twice in passenger planes and compared the clocks' times with those of identical clocks on the ground. The travelling clocks had slowed, if only by a tiny fraction.

This strange dilation of time means space travellers could become time travellers, Mallett says, journeying vast distances but ageing far fewer years than those awaiting their return on Earth. 'Your heart is a clock, it will keep time for you, and it will slow down, too, once you hit high enough speeds,' Mallett says.

He points to the classic film *Planet of the Apes*, where our heroes land thousands of years into the future on an Earth ruled by super-intelligent apes, after a lightspeed space flight of just two years. To get to some of the distant planets now being studied through telescopes by astronomers, space flight will have to get much, much faster.

But nothing can break the speed of light. Photons or light particles can travel this fast because they have no mass – at light speed, time itself falls away and the journey seems instantaneous. But for something with mass, such as a spaceship, to travel that fast would require an infinite (and so impossible) amount of energy. And our technology is still a long way off moving humans anywhere close to light speed.

'But who knows what we will build in the future,' Mallett says. 'If we need to move faster through space, perhaps we could design a way to move space itself, instead of just us.'

In the subatomic world, where there is considerably less baggage slowing things down, scientists can already accelerate particles hair-raisingly close to light speed. For a tiny proton in the sprawling Large Hadron Collider beneath the French–Swiss border, for example, 11 months of Earth time will seem like just one second.

Mass also distorts spacetime, like putting a bowling ball on a trampoline. We call this effect gravity. Out in space, where the gravitational pull of the Earth is less strong, time speeds up. It's why the GPS satellites that guide our smartphones and cars on the ground have to be synced up with Earth clocks roughly every few minutes to stay accurate – otherwise they run fast by 38 microseconds each day.

But in a black hole, where gravity is so intense even light cannot escape, time might appear to stand still. Some scientists wonder if travelling near, but not into, a black hole could act as a 'natural' time machine, slowing time for those making the journey – 'yes, like the movie *Interstellar*', Mallett says. In that 2014 science-fiction epic, astronauts escaping a dystopian Earth use the high gravity of a black hole to slingshot through time to faraway planets.

'And if you have a rotating black hole, which we sometimes see, then it doesn't just bend spacetime, it twists it, like stirring a coffee with a spoon. Our Earth rotates and has gravity too but a black hole has so much gravity it can twist both space and time. That could create not just a way forward but a way back – a loop.'

Could we actually travel back to the past?

On 28 June 2009 Stephen Hawking threw a party for time travellers – and no one came. The sly scientist had posted an open invitation the day after the soiree, meaning only those already in the future with a means to travel back into the past could have attended.

So does Hawking's disappointing turnout mean time travel never becomes a reality?

It's not a definitive no, but many experts say there is an inherent problem with the logic of backwards time travel, regardless of the science.

Hawking himself said it must be impossible. To travel back in time would invite the possibility of changing the past, which doesn't make logical sense seeing as it's already happened. He supposed there must therefore be something about the universe we still don't understand – probably in the strange world of quantum mechanics where subatomic particles seem to defy the usual physics – that would render backwards time travel impossible.

But in his last book, published posthumously in 2018, Hawking conceded that time travel remained a very serious question.

To travel back in time would invite the possibility of changing the past, which doesn't make logical sense seeing as it's already happened.

Here's the problem: travelling into the future doesn't break the direction of reality – time (and causality) moves forward as one moment leads to the next. But to go back, Miller says, you'd have to turn that around 'so what you do in the next minute causes you to go backward, affecting things in reverse. The tea on the counter goes from cold to hot, from being left out to just poured.' (Director Christopher Nolan's blockbuster film *Tenet* centres around this strange idea of time reversal.)

One potential solution again lies in Einstein's mathematics. He theorised it was possible to fold spacetime, creating a tunnel between two distant locations, a shortcut known as a wormhole. Each endpoint could be a vast distance apart in both space and time, Miller says, 'punching a hole through spacetime', making the concept a convenient staple of science fiction from *Star Trek* to *Stargate*.

As you read this, scientists are hunting for wormholes out in the universe with radio telescopes, but if we find one or even figure out a way to conjure up our own, Miller says, we still don't know if we will be able to travel through them – or where we'd end up.

To hold the mouth of a wormhole open in spacetime, say, long enough for a spaceship to travel through, we'd need a special kind of force, an energy without matter known as negative or dark energy. Since the Big Bang, the universe has been expanding, but that growth is not slowing, as one would expect billions of years after the initial explosion – it's accelerating. Scientists think a mysterious force known as dark energy is driving it.

Mallett, meanwhile, believes he's found another theoretical way to fold spacetime – using light. 'Einstein showed that light produces a gravitational field too, so if light has gravity and gravity affects time then light could affect time.'

Mallett theorises that a circulating beam of light or 'ring laser' with enough power could twist not only space but time as well. 'Kind of like our rotating black hole and stirring that coffee cup,' he says. With the right geometry, he thinks it could even fold back into a loop, allowing travel into the past.

Demonstrating this beyond maths equations would require serious funding, he concedes, but as with all such experiments in spacetime, it could lead to beneficial spin-off technologies.

'I'm looking more realistically at sending information back in time rather than people, which would require a [gigantic] amount of energy if it was proven to work,' Mallett says. 'But if we could send subatomic particles back like neutrons, they have an "up" and a "down" state we could turn into binary code [and] we could warn people in the past about what was coming. We could save lives.'

He recalls a letter he once received written in German. He doesn't speak the language but from the accompanying photographs, he could understand the story: in the first photo a young woman was pictured with her family, in the second was a grisly car wreck. When Mallett had the letter translated, he learnt it was from the woman's father, who wanted to know if he could send a message back in time to himself on the day of his daughter's fatal car crash to warn him of what was to come.

So what does all this mean for would-be time travellers?

Mallett agrees the twisty logic of backwards time travel is difficult to reconcile. But he notes the subatomic world – where particles can be in two places at once – may yet offer answers.

'If science has taught us anything it's that nature is weird,' he says. 'The question is open and it's no longer taboo. It wasn't until this century I could come out of the time travel closet, as it was, and reveal I'd been researching it, but there's lots of people looking at it now.'

Still, if he does build his laser time machine, Mallett admits it will never fulfil the wish of his ten-year-old self.

'We would only go back as far as time has been bending, when the machine was turned on and the loop started,' he says. 'I wouldn't be able to go back to my father.'

Unlike using natural phenomena such as a wormhole to time travel, a machine built by humans would be expected to come with this inbuilt limit: you could travel only as far into the past as the machine had been running. This perhaps explains, Mallett says, why we're not already inundated with time travellers from the future. 'Year zero' for time travel – marking the beginning of possible journeys – is yet to arrive.

Perhaps we will work out how to use black holes and wormholes after all, where the scale of the past they could unlock would be staggering.

But he's not ruling out anything. Perhaps we will work out how to use black holes and wormholes after all, where the scale of the past they could unlock would be staggering. Perhaps an advanced alien race will reveal itself with its own (long-running) time machine or a way around time, as is imagined in the film *Arrival* where (spoiler alert) aliens visit Earth with the 'gift' of altered time perception.

Or perhaps the answer is already etched in the logic, as Hawking guessed, a painful truth of the universe waiting to be confirmed. No matter how far science may take us, we can never go back.

CAN WE CHANGE THE PAST?
A dialogue on time and space

Some suppose there's no way to return to the past and *not* change it. Huge consequences could follow from simply breathing (or from trampling on butterflies, as in the butterfly effect, first imagined in Ray Bradbury's story about time tourists, *A Sound of Thunder*). But most experts say the logic tells a different story. In this edited conversation with reporter Sherryn Groch, philosopher Kristie Miller explains:

SG: Okay, Kristie, suppose I've built a time machine and I travel back to 1945, before my parents were even conceived, to see my grandfather.

KM: Ah, the Grandfather Paradox.

SG: Shhh, act surprised. So, suppose I hate my grandfather and this whole time travel jaunt is actually an assassination. Wouldn't I, in killing my poor old grandfather, also wipe my father and myself out of existence, thereby leaving no one to go back in time to do away with the old man in the first place?

KM: Aha! But, you see, you'd fail to kill your grandfather!

SG: But who's going to stop me? You?

KM: No, you're right. Most philosophers and physicists agree there's no time travel guardian or special force preventing you from doing anything back there. But you do exist and your father exists (and granddad too) so that tells us you didn't kill him.

SG: But I really went back in time. I took a selfie. Doesn't just being in the past change it?

KM: From a God-eye perspective looking down at the whole block, all of time, I'd see you get out of your time machine in 1945, maybe crush a few butterflies or accidentally knock over an old lady on your way to find your grandfather, break her hip. You've already made heaps of changes, right? But 1945 doesn't play out once without you and then again once you've cracked the secret of time travel decades later. If you ever manage to get to the past then that means you've always been part of it. You always bowled over the old lady. Your time machine is just an explanation of how.

SG: In the movie *Back to the Future*, when Marty McFly helps his parents get together, he was always the reason that happened?

KM: Yes.

SG: So *Back to the Future* could happen?

KM: Nice try.

SG: But if I was already there in 1945, why don't I have any memories of this homicidal rampage in the present?

KM: Because in your personal time – the time you experience living one moment to the next – you haven't yet built the time machine and gone back. In external time, the block, 1945 comes first but for you all this would happen in the same linear order. You build the machine, you hop in and return to the year 1945, where you make those memories.

SG: Wouldn't my grandfather remember what I did? I left him alive to blab to the cops, after all.

KM: Yes, or at least he'd remember what he experienced back then, that in 1945 a young woman appeared hell-bent on murdering him. Perhaps he wouldn't recognise it was you until you'd grown up. Perhaps this is why you have such an objectionable relationship in the first place.

SG: But what if he did remember and tried to stop me being born?

KM: The same rule applies to him – you were born so whatever he did, he'd fail.

SG: Hold on, you can't change the future either? Does that mean I'm locked into going back? What if my grandfather finally worked up the courage to confront me about what I did and I changed my mind?

KM: Sure, you could have even taken a vow to never go back. But if the past says you went back, if you made that decision at one point, then you did go back. There's no replay. So something would happen to send you back. Perhaps while cleaning your time machine you accidentally start it up, and remember something to reignite your homicidal rage at your grandfather.

SG: This all sounds a lot like fate. What about free will?

KM: When we think of the future as fixed, it can feel like destiny because of the way we see time but the past is fixed too, isn't it? You can't go back and change what you ate for breakfast.

SG: Well, not if you've already ruined my grandpa-murdering fun.

KM: But no one forced you to have Corn Flakes, right? That wasn't the will of some divine Corn Flakes god. You made a free decision in the past and you made one in the future. But it's fixed in the sense that the same moment doesn't appear in spacetime multiple times over.

SG: What if time travelling splits me off into a parallel universe where I did kill my grandfather?

KM: If you left your universe, you'd be changing the past of a different world, not your own.

SG: I think I need to lie down.

23

HOW DO YOU SOLVE HOMELESSNESS?

Economists and social workers agree
homelessness can be fixed. But how?
Is it really as simple as building more homes?

Sherryn Groch

Bob Petersen slept on trains for two years before the pandemic. Nearly every night, he'd ride more than 100 kilometres of rail, from Sydney to Kiama and back again. Sometimes he'd wake in Newcastle, sometimes Port Kembla, but he always paid his fare. 'And I tried to keep myself neat and tidy so no one would bother me,' the 74-year-old says.

Now he has a home.

In Melbourne, Glenn Kent was also excited to pay his first month's rent in the winter of 2020. After more than 15 years sleeping rough, and a lifetime bounced around clinics, shelters, jail and then COVID hotels, Kent, 50, finally got the keys to a home – a unit with rent costing just a quarter of his disability pension, somewhere he can fix up his bike, bake cakes for the local church, return to his artwork and 'keep out of trouble'.

This is the answer to homelessness: more affordable homes. It might sound too simple. Or, as Kent first thought, too good to be true. But if the pandemic has taught us anything, it's that much of what was once considered impossible, or at least impractical, can be achieved, and quickly, if we want it enough.

As the streets of our cities emptied to stop the spread of COVID-19 in 2020, tens of thousands of Australians without shelter were put up in hotels and rentals as part of a mammoth effort by state governments and frontline services. For a lucky minority, such as Petersen and Kent, the new digs became permanent. But many have already found themselves back where they started.

'People walk past homelessness thinking it's one of those things that'll never change,' says University of Queensland researcher Cameron Parsell. 'They also think it doesn't cost them anything. They're wrong on [both fronts].' Parsell's work has shown that leaving someone on the street long term, racking up otherwise preventable trips to emergency departments, shelters and watchhouses, costs about $13,000 more per person every year than placing them in supported housing.

During the pandemic, with the economy in need of a jobs-intensive stimulus and thousands more Australians facing homelessness, economists say the business case for building more of this social housing has become stronger than ever.

'We started with a housing crisis in Australia and it became a health crisis, which became an economic crisis,' says Bevan Warner who heads the frontline homelessness service Launch Housing in Melbourne. 'Now we have an opportunity to solve all three.'

So how much housing do we need and what would it cost? And how have countries such as Finland radically reduced homelessness already?

How common is homelessness?

No one knows exactly how many people are homeless in Australia. For every person sleeping out, dossing down in doorways and parks, many others are crushed into overcrowded shelters and boarding houses, or put up on couches and in garages. On census night in 2016, more than 116,000 Australians were recorded without secure housing or the prospect of finding it, 8200 of them sleeping rough. But when the pandemic hit our shores in 2020, the true scale of the problem became clearer than ever. About 40,000 people were put up in hotels in Victoria, New South Wales, South Australia and Queensland as part of an emergency COVID response between March and September, vastly outstripping existing rough-sleeper counts. Researchers say the real number housed is likely even greater.

'We may never know the full story,' says UNSW Professor Hal Pawson. 'It's amazing what [was] done so fast. Of course, it wasn't about ending homelessness, it was about stopping the virus spreading.'

Warner says that, while the rapid response of state governments deserves credit, many of those housed during the crisis had already been churned in and out of support services for years (in Melbourne, an average of eight years). Victorian government statistics show more than 70 per cent of people assisted by homeless services in 2018–19 were still homeless at the end of support, for example. 'Clearly, the system hasn't solved their homelessness,' Warner says. 'We've just been treating the symptoms, not curing it.'

Why are people homeless?

On the street, people offer familiar reasons for how they got there, the tragedies and misfortunes that seemed to hit all at once: deaths and relationship breakdowns, job losses, car crashes, abuse. Family violence is now the leading problem pushing people, many of them children, out of their homes. Mental illness and disability are also common, and Indigenous Australians are more likely to find themselves without a roof over their head.

For Petersen, it began with the sudden death of his long-time partner. 'I'd been working in a hotel but I had to move away after that,' he says. 'I was devastated, and broke from raising the money for her funeral.' After renting with a friend didn't work out, Petersen jumped on a train. 'I just thought I'd travel until I could figure out what to do.' But he didn't find anywhere he could afford to rent by himself and those nights on the train turned into years.

For other rough-sleeping regulars such as Kent, who has struggled with serious mental illness and addiction as well as literacy issues due to dyslexia, housing came with too many hoops to jump through. Getting clean on the street was too hard, shelters were too cramped, pets too precious to give up for a temporary bed.

'I used to get work, I was a pastry chef and a landscaper, then it got harder,' Kent says. 'When you're sleeping in public toilets, and you haven't had a shower in two weeks, you're out of jail, no one will give you a go. I couldn't even fill out the forms with my name and address [because of] my dyslexia. I couldn't sleep. So I used drugs to stay awake on the street.'

Homelessness doesn't happen to a certain kind of person. Many are just on the wrong side of Australia's housing boom, born without a social safety net or priced out of increasingly competitive rental markets. Men are more visible on the street, and older women are now the fastest-growing demographic in homelessness statistics, but young families – women and children – still make up the bulk of those without secure housing. The numbers are likely even greater than we think, experts say, with some women returning to violent homes to spare their children nights sleeping rough. Kids on their own, meanwhile, can be missed in the data altogether, too young to get a bed in an adult shelter but often wary of ending up under the nose of child protection services.

Even before the pandemic, people were in increasingly precarious positions, says Michele Adair, the chief executive of the not-for-profit Housing Trust, which helped Petersen. 'People don't know this but when I divorced years back, I had two little kids and we came very close ourselves,' Adair says. 'Now COVID seems to have woken [people] up. It's always been someone else doing it tough, just addicts and no-hopers, now suddenly it's their mum, their brother, their friend in trouble. We have teachers and nurses, essential workers, in serious housing stress. This is everyone's problem.'

Even before the pandemic, people were in increasingly precarious positions.

In recent years, as rough sleeping, or street homelessness, has become more visible in our cities, it's hit an 'embarrassment threshold' for governments, too, Parsell says, triggering more spending on crisis interventions such as street outreach. But there are not enough places for people to go.

Homelessness doesn't happen to a certain kind of person.

Shelters are almost always full, and women's refuges overstretched. Welfare support to help low-income earners rent privately has not kept pace with the soaring cost of living. Meanwhile, public or social housing, which offers secure tenancies to those most in need at about 25 per cent of their income, is declining, which has blown out wait times into years, even decades. As of June 2020, there were almost 160,000 families in that queue around the country.

Until recently, mother of six Faith Labaro was one of them. She recalls couch-surfing with young children to escape a violent partner. 'We jumped on a train in the middle of the night to a friend's,' she says. 'But we'd been on the public housing waitlist for six years.' Labaro was forced to return to her partner for a time as she struggled to find an affordable rental, often scraping by with food donations from local charities before Housing Trust found the family a social housing unit in early 2020.

Decades of belt-tightening has seen public housing in effect halved since it reached its peak around the 1990s, says Pawson. Just 4 per cent of all homes in the country (or about 396,000) are government stock – compared to more than 15 per cent in the United Kingdom, 20 per cent in Denmark and about 25 per cent in countries such as Scotland and Finland. Victoria has the lowest public housing supply in the country per head, and South Australia the most.

What's the single best solution?

There's no one easy fix to homelessness because there's no one cause, but when we follow people like Petersen and Kent through the system, research shows one pathway does have a better ending than most. It's called Housing First and it's exactly how it sounds. Instead of the traditional 'staircase' model where long-term housing is 'earnt' by meeting certain conditions, Housing First offers it straight off the bat – without the usual sobriety or income tests, the long queues, the paperwork. Once a need is identified, the first priority is finding that person a stable place to live – addressing their other issues, such as addictions, can come next. It was first tried in the 1990s in New York by psychiatrist Sam Tsemberis, who had begun recognising his former patients sleeping on the street and realised many people just weren't making it through the system. Where Housing First has been rolled out since, including Scotland, Canada, and Scandinavia, people are more likely to turn their lives around.

Adair recalls a nurse fleeing domestic violence in New South Wales who had to sleep in her car and in hotels for almost three years before Housing Trust found her somewhere permanent. 'Sometimes you can't wait for the paperwork,' Adair says. 'You can't settle your kids, you can't get a job, you can't do anything until you have a roof over your head.'

Those who find themselves homeless long term are likely to die 15 to 30 years earlier than the rest of us and are often plagued by preventable medical problems. Most people fronting up to support services don't have this kind of chronic homelessness, but those who do take up a fair chunk of the sector's limited resources. So, as Adair explains, in systems where you solve this long-term homelessness first, the benefits (to both lives and bottom lines) go up and the bottleneck in shelters and social housing opens.

*Those who find themselves homeless
long term are likely to die 15 to 30 years earlier
than the rest of us and are often plagued
by preventable medical problems.*

Consider the case of US veteran 'Million-dollar Murray' Barr, documented by writer Malcolm Gladwell in *The New Yorker* in

2006. Local cops spent so much time taking Murray to hospital that it would have been cheaper to put him up in a hotel with his own nurse. When he was sober, Murray was a talented chef; smart, charming and beloved by the nurses and cops who knew him. But by the time he died of intestinal bleeding, it was calculated that he'd cost more than a million dollars over a decade on the street.

Where Housing First has worked well, it's been enshrined into national policy, says University of Queensland researcher Andrew Clarke. 'In Finland, housing guides everything.' The country spent 250 million euros ($418 million) building 3500 new homes and hiring 300 extra support workers – and is estimated to save about 15,000 euros ($25,000) a year for every person housed. Homelessness is down more than 35 per cent since the Housing First plan began in 2008, and long-term homelessness has more than halved. In the capital Helsinki, just one shelter is needed during winter.

Australia's COVID hotel response has shown that Housing First can work on a big scale here, too, says Parsell, even if most housing offered so far has only been temporary. 'NGOs told us [in 2020] that for the first time that they could remember they were able to get every person they saw a bed, and people who wouldn't normally take offers, the ones wary of being shuttled through shelters and back out again, they were saying yes, too. 'There's been successful pilots in most capital cities in Australia before. The hard bit is scaling it up – and getting those support services in place to help them rebuild their lives.'

During COVID, some states have had a real crack at housing people permanently while they are off the streets. Both New South Wales and Victoria have committed tens of millions of dollars to move a couple of thousand people from hotels into rentals, including caseworker support. Short-term, they hope to get around the social housing shortfall by head-leasing: taking up the lease of private properties and paying the gap between market rent and what someone in need can afford. But both state schemes last only two years and have moved slower than hoped.

The United Kingdom ran a similar transition program during the pandemic and by early 2021 it had housed two-thirds of the 30,000 people it put up in hotels. In Australia, about 8000 rough sleepers had left emergency COVID hotels by September 2020 but only a third of those had been helped into longer-term accommodation.

Kent, like many of the people put up in hotels during the pandemic, was at once grateful and wary. Surely it couldn't last forever? Now, since moving into his long-term rental in Melbourne with help from Launch, he's managed to kick drugs for the first time in decades. 'Things taste better. I have my own pillow. I can actually sleep.'

Of course, there are things Housing First can't fix, as founder Dr Tsemberis himself has acknowledged. It can't fix a mental health system that's failing the vulnerable, or a housing market stacked in favour of wealthy investors. It can't fix poverty itself. And, while Australian experts agree building more social housing is the best solution in the long run, many frontline workers also stress that more crisis beds would help people sleeping rough in the meantime.

Is social housing good for the economy?

Social housing requires a huge upfront investment in bricks and mortar, plus running costs – numbers that have long made both sides of government pale. But a big building blitz also creates jobs.

University of Melbourne economist Lisa Cameron lives in what was once a public housing unit, built during the Second World War, and says the business case for creating more public housing today is as strong today as it was in 1945. Again, the economy needs a big stimulus spend, this time as it reels from pandemic lockdowns. And, with the bond rate low, she says the federal government can borrow money for very little interest. And then there are the flow-on benefits – not just jobs for tradies but more people back on their feet and contributing to the economy.

When the Economic Society of Australia polled 49 top economists on their wish list for COVID recovery spending, social housing topped the list (just ahead of permanently boosting welfare payments). But it was absent on budget night in 2020. Instead, Treasury earmarked $1 billion worth of cheap loans for community providers to help them build more affordable housing – that is, private rentals generally offered around 20 per cent below market rent. At the time, the Coalition acknowledged that more social housing might also be in order but insisted the money should instead come from the states, though both levels of government fund it. In 2021, that position hadn't changed (but more money was put into frontline domestic violence services).

Clarke notes that the initial heyday of public housing after the Second World War was driven by the Commonwealth, as were resurgences during the 1970s and then, briefly, during the Global Financial Crisis with a stimulus splash by the Rudd government. 'It's always come from the federal government, they have the deepest pockets for revenue,' he says. 'The states can't do it alone.'

Federal Labor shied away from promising more public housing at the 2019 election but in May 2021 it pledged to build 20,000 new properties in five years if elected, 4000 of them specifically for women and children fleeing violence as well as older women on low incomes. This new $10 billion fund would create another 10,000 affordable housing properties for frontline workers, and generate an expected 21,500 jobs each year. Labor leader Anthony Albanese spoke of a rare chance to reinvent Australia's economy post-COVID as he unveiled the plan, adding: 'I grew up in a council house in Camperdown, the only son of a single mum on the disability pension. I'll never forget that I'm here tonight because good government changed my life.'

The proposal has been praised as a decent 'first step' by the homelessness sector, even if it has been dwarfed a little in ambition by Victoria's own big build. In late 2020, the state announced it would create 12,000 new public housing units in four years. That will cost $5.3 billion but is expected to generate $6.7 billion in economic activity, including more than 10,000 jobs a year.

Grattan Institute economist Brendan Coates says a national social housing build would offer more construction jobs than the transport infrastructure projects so far at the centre of the Morrison government's COVID stimulus, and would boost the economy more than tax breaks. While Australia's economic fortunes are recovering faster than expected, Coates says there's still a case for further stimulus into 2021 and beyond. And, at the end of a big social housing spend, Parsell notes, we'd be left with something even more meaningful than economic growth. 'You will have changed lives. And the country will have a housing safety net.'

But how much would all this cost?

Estimates for how many more social housing units Australia would need range from the tens to the hundreds of thousands. Those numbers may have changed a little under falling migration during COVID, but Pawson says more people are also finding themselves in poverty for the first time as pandemic eviction bans and income support have wound down. And that has created a stampede in the cheaper end of the rental market, Warner says.

Some experts would like to see Australia ramp up its social housing stock more in line with countries such as the United Kingdom at 15 per cent of all dwellings. But Coates says even growing the stock back up to 5 per cent from 4 would mean building about 100,000 new homes in total, which he estimates would set the government back about $30 billion.

'That's as large a number as I can imagine a government going in one hit, even one committed to doing something big,' he says. 'We'd recommend building 30,000 homes for about $10 billion to start.'

While there are clear savings to offering housing to the most vulnerable, Coates says value for money fades once you start putting those further up the ladder into social housing. 'I don't think everyone on a low income needs social housing. Early intervention and raising Renter's Assistance and [other welfare] is a more effective way of targeting those people who can stay in the private market with a bit of help.'

Some countries that have rolled out Housing First widely, such as Scotland, already had high levels of public housing to begin with. 'They have billions of dollars of these assets,' Coates says. 'For us to build or buy up that much stock would be hugely expensive. Singapore has great housing as well but the government owns 80 per cent of it.'

There are cautionary tales, too. In New Zealand, Prime Minister Jacinda Ardern's ambitious plan to build 100,000 affordable houses in ten years was abandoned after just two, miles behind its targets. Many of the homes had been built in areas where no one wanted to live. The Bush administration also failed to plan appropriately when it was trying to scale up Housing First across the United States – many houses were not directed at the people most in need, and so did not make up enough savings.

What works in one country can't simply be copied, says Pawson. 'What Finland did is not a magic bullet. You need to adapt it.' And big interventions may require changes to the law, too, not just budgets. An enshrined right to shelter is coming into vogue in some parts of the world where Housing First is scaled up, including Finland, Scotland, France and Canada. In other places, such as in the United Kingdom and Europe, local authorities must ensure people have alternative housing before they can be evicted.

Could private investment save the day?

The private sector is already working to tackle homelessness, in its way. But, while clever plans to repurpose unused office space, car parks and even buses as accommodation are welcome, experts say the market alone won't solve homelessness, at least not long-term. A big housing build needs big government buy-in.

Still, Australia could offer tax incentives for private enterprise to invest in more affordable housing or social housing run by charities – as happens in countries such as the United States with a long history of philanthropy. A mix of public and affordable housing can make developments more appealing to investors, including governments, and stop concentrating disadvantage in entire neighbourhoods. About $13 billion in tax credits is given out each year via negative gearing to help Australians buy a second or third home, so, Warner asks, 'why isn't there anything for people without one?'

But Coates offers a warning from our not-too-distant history. When the Rudd government invested in housing to stimulate the economy after the GFC, it launched two main programs. The first spent $5.2 billion to build 20,000 social housing units and refurbish 80,000 others in two years and, after coming in on time and under budget, was heralded a success. The second, which offered tax incentives to private enterprises to build affordable housing, cost $3.1 billion but was abandoned as a failure in 2014.

The problem? Coates says the government handed out too much money to the market for what it could have essentially done itself at half the cost.

Affordable housing is cheaper than social housing. In the case of the Rudd-era scheme, that should have cost the government just $4000 a year per household but instead, a Grattan Institute analysis found, it was paying $11,000 – almost as much as public housing. It was, in effect, a $7000 windfall for the private sector.

Still, while solutions sometimes struggle to find purchase

at a national level, cities around the world are making a real difference at street level. Eleven cities in the United States have reached a standard called 'functional zero', meaning that when homelessness does happen it's rare and it's brief. Adelaide and now, to some extent, Sydney are aiming for the same target, working to improve street counts and coordinate their data across support services. The idea is to know everyone on the street by name and circumstance, so resources can be targeted.

At Launch, Warner hopes to turbocharge the approach, too, with a Melbourne Zero plan: 'It's about [using] real-time data to respond to what we're seeing on the street. It's about not walking by any more. It's about knowing their names.'

Kent, Petersen and Labaro still find it hard to describe what it means to be known in their communities again, to have found a home while the world around them was turned upside down by a pandemic. Kent is fixing up his bike as he gets his fitness back, hoping to raise funds one day soon at a charity ride for homelessness. Labaro is studying graphic design and her kids are thriving at a nearby school. Petersen, meanwhile, speaks highly of the casseroles his neighbours drop in.

'It's changed everything,' he says. 'At first it was like, "I can breathe again." Now it's, "Well, I'm living again."'

24

WHY IS BREAKDANCING IN THE OLYMPICS?

Head spins and air flares will be de rigueur
in Paris in 2024. What does it take for a
new sport to make it into the Olympics?
And how does a B-Girl or B-Boy win gold?

Iain Payten and Sarah Keoghan

It is a pastime we usually associate with New York in the 1980s, Run DMC and that guy from accounts doing the worm after a few drinks at the work Christmas party. But breakdancing is now not only a famous hip-hop dance style, it is a sport. And not just any sport, it's an Olympic sport – following a bold decision by the International Olympic Committee to have B-Boys and B-Girls compete at the 2024 Paris Games.

The inclusion of breaking has come as part of a push by the International Olympic Committee (IOC) over the past decade to modernise itself and to appeal more to a youthful audience. The IOC had observed the average age of audiences for Olympics creeping up, and decided to open the doors to events such as surfing, skateboarding, 3x3 basketball, rock climbing and breaking.

Most of the usual Olympic staples remain – there are about 25 core sports, from athletics and swimming to gymnastics and equestrian – but now they have younger, noisier neighbours. And not everyone is happy. Many traditionalists are aghast, and debate rages about whether breaking is a sport, let alone an Olympic one. Having seen her sport miss out on an Olympic spot during the IOC's four-yearly review process, Australian squash star Michelle Martin said the decision made a mockery of the Olympics.

> *Many traditionalists are aghast, and debate rages about whether breaking is a sport, let alone an Olympic one.*

So how did a dance style born on the streets make it all the way to the five-ring circus? What are the attributes of the athletes involved? And, when it comes to deciding a gold medal, will it be a good old-fashioned dance-off?

First, what is breakdancing?

A small point of order to start. The correct term is 'breaking' and, since an IOC vote in December 2020, there will be Olympic gold medals in 2024 won by the world's best breaking dancers.

Breaking is one of the most unusual sports ever added to the Olympics program, given it has always been considered mostly edgy artistic expression (although an argument could also be made for artistic, formerly known as synchronised, swimming). The dance style was a child of the hip-hop subculture that emerged on the streets of the Bronx in New York in the early 1970s. DJs figured out that the most popular sections of disco and funk tracks were the breaks between the verses – no singing, just music. So they looped, or repeated, these 'breaks', stretched them out and had rappers voice rhymes over the resulting tracks. The drum-based music is perfect for breaking routines, which can combine fancy 'toprock' footwork, various types of 'windmills', 'flares' and 'swipes', head spins, shoulder spins, back spins, all sorts of acrobatic flips – and 'freezes', in gravity-defying poses.

The dance style was a child of the hip-hop subculture that emerged on the streets of the Bronx in New York in the early 1970s.

Breaking became huge in the United States and around the world in the 1980s. Leading groups such as the Rock Steady Crew performed for presidents and Queen Elizabeth. Famous actor Vin Diesel hilariously was found to have starred in an 80s instructional video on how to breakdance while trying to break into Hollywood as a teenager. Breaking fell out of wider popular culture but has remained a foundational pillar of hip-hop culture alongside DJing and rapping.

Nice, but how is 'breaking' now an Olympic sport?

The first Olympic Games were held in Greece in 776BC. They consisted of but one race – over 192 metres, called the 'stade', whence 'stadium'. Athletes from rival city-states and kingdoms would compete at the sanctuary of Zeus in Olympia every four years. Over the centuries more sports were added, including long jump, equestrian, discus, pentathlon and boxing, which have carried through to the modern era. (Chariot racing and a brutal boxing-wrestling hybrid called pankration, in which only biting and gouging were against the rules, were left in the past.) Winners were awarded laurel wreaths. The ancient Games continued until 394AD when the Christian emperor Theodosius banned 'pagan' festivals. Irregular 'Olympic' competitions began to be held again from the seventeenth century in Europe but it wasn't until 1894 that French historian Baron Pierre de Coubertin formed an International Olympic Committee and helped stage the first modern Olympic Games in Panathenaic Stadium in Athens in 1896.

In recent years the IOC began worrying the Olympics were seen as boring, stuffy and irrelevant, particularly to young people. 'Let's face it: the Olympics is old, man,' wrote Clio Chang in *The New Republic* in 2016, noting the median age of US viewers for the 2008 Beijing Olympics was 47, rising to 48 for the 2012 London Games. Fearing the whole spectacle would slide into the grave right after its supporter base, IOC president Thomas Bach said the Olympics had to 'change, or be changed'.

In recent years the IOC began worrying the Olympics were seen as boring, stuffy and irrelevant, particularly to young people.

So the IOC began looking to make the Games more youth-oriented, trading the classic events for things twenty-first-century youngsters actually do in their spare time. It had been successful in steering towards youth in the Winter Olympic world, when snowboarding was added to the 1998 Nagano Games. 'We want to take sport to the youth,' Bach said in 2015. 'With the many options that young people have, we cannot expect any more that they will come automatically to us. We have to go to them.'

In 2016, the IOC announced an unprecedented shake-up. After inviting submissions from sporting federations around the world, surfing, rock climbing, skateboarding and 3x3 basketball were chosen to be included on the 2020 Olympics schedule in Tokyo. The IOC further decided to give future Olympic hosts the ability to propose sports they believed would work in their Games. Tokyo organisers put forward baseball, softball and karate, which are hugely popular in Japan, and those sports were also added, although they will not be part of the 2024 Games.

The French, meanwhile, considered proposing parkour, a kind of fast-paced running, jumping and climbing over city structures that has made it into James Bond films. But the people behind Paris 2024 instead pushed for breaking. It will be staged outside at Place de la Concorde, at the end of the Champs-Élysées.

And it's not as if there is no precedent for an odd and short-lived Olympic sport. Tug-of-war, horse long jump and live pigeon shooting are among the stranger – albeit popular at the time – sports to have been tried and discarded at the modern Olympics.

Breaking famously caught on in the banlieues (suburbs) of French cities, and France is now reportedly the second biggest hip-hop market in the world after the US. Melbourne choreographer Antony Hamilton, who runs the Chunky Move dance studio, says breaking began to develop in Europe and the United Kingdom shortly after hip-hop took off in New York City. 'The people who were in the hip-hop scene in New York in the early 80s were being promoted by art dealers and they were being flown all over the world by promoters like Malcolm McLaren who was famous for commercialising hip-hop,' Hamilton says. 'So, you had this intercontinental explosion of hip-hop as an art form and because of the close proximity of the east coast to the UK and France . . . it was a quick germinating cultural movement.'

Hamilton says the ability for children of all walks of life to go out on the streets and learn breaking has led to it transcending regions and class boundaries. 'The grassroots nature of breaking will always be an important part of the culture and an important part of the learning. If it's not within the confines of the cultural practice of the street and communities . . . it loses its authenticity,' he says.

Another possible reason the IOC greenlit the addition of breaking is that hip-hop is growing at a rapid rate in China. A reality TV rap program there in 2017 caused the DJing, rapping and breaking scene to explode in popularity and a recent estimate counted the 'street dance' community at more than 10 million.

How does any new sport make it into the Olympics?

The IOC is one of the world's great bureaucracies so, unsurprisingly, there is a rigid formula applied in choosing new Olympic sports. Apart from a host nation nominating a chosen sport (which has to be approved by the IOC), sporting federations are invited to apply for inclusion when the Olympics reviews its program every four years. Some make it, most don't. And when there are new sports coming in, that means an event has to be discarded. As with the Olympics themselves, this process has a few winners and many more unhappy losers.

There was a rule from 1972 that a sport must be widely practised on four continents, in at least 75 countries for a men's sport and 40 for women, to make the Olympics cut, but this was ditched in 2007. Today, there are five basic principles that the IOC weighs in deciding a sport's suitability. The first is termed Olympic Proposal, which includes the sport's history, whether it has been in the Olympics before, the size and scale of the sport internationally. The second looks at the sport's Institutional Matters – that is, financial status, governance and gender equity. The capacity for the sport to be played by men and women, or in mixed teams, is now one of the IOC's major considerations. The sport's image and alignment with Olympic values is the third factor, and fourth is its popularity: how many people will watch it, how much media it will attract, how many sponsors will back it and whether the world's best athletes in it will take part in the Olympics. The last category is a sport's business model – how much it costs to stage and how much income it makes.

Seasoned Olympic experts say the key factors are appeal to a young audience and global popularity. The IOC, unsurprisingly, wants sports that will attract ticket-buying spectators, media interest, television rights and sponsors – that's why golf and rugby sevens were added in Rio ahead of squash, which campaigned heavily.

So it makes sense that three-on-three basketball would have been embraced for Tokyo: a shortened version of basketball, played on a half-court, 3x3 (pronounced 'ex') is aimed at audiences who play impromptu games in neighbourhoods around the world. For the Olympics, it's designed to be short, explosive and friendly for those with short-attention spans, as is Twenty20 cricket (which is a hot tip for inclusion at Brisbane's 2032 Games). Cricket was incidentally, a sport at the 1900 Olympics, but only two nations entered teams – host France and Great Britain, represented by Devon & Somerset Wanderers Cricket Club. Perhaps not surprisingly, Great Britain won.

And both men's and women's surfing tours are also hugely popular with not only fans but media and sponsors, making it a no-brainer to add, starting with Tokyo. The same goes for another

The key factors are appeal to a young audience and global popularity.

Tokyo Games event, sport climbing, which is extremely popular in the United States and Europe – at the Olympic level, it's frantic, spidermonkey stuff. And the kite-foil surfing event that will debut in 2024 is another extreme yet accessible sport that can be practised by aspirants at any location where there's open water.

What's a B-Girl or B-Boy judged on?

The breaking format for the 2024 Olympics is almost certain to be based on good old-fashioned dance battles. That's how it worked when breaking was a trial event at the 2018 Youth Olympics in Argentina, and an Olympic format and judging criteria had to be created from scratch.

Competitors performed in showcase rounds before the best 32 went to battle rounds. After a series of six turns each on the stage – over a battle lasting five to six minutes – the judges, who included Richard 'Crazy Legs' Colon, leader of the aforementioned Rock Steady Crew, assessed the combatants' performance. And so the knockouts continue until there is only one left – the gold medallist.

As with other dance sports – think, *Strictly Ballroom* – breaking is judged subjectively by a panel. In Argentina, from the array of accepted moves and styles that leading dancers already use, veteran breaking experts created a system that assessed three categories: Body/Physical was based on technique and variety; Soul/Interpretative was about quality, composed of 'performativity' and 'musicality'; and Mind/Artistic assessed creativity and personality. It is a compilation of broad categories that mirror other dance sports, and sports with subjective judging such as diving and gymnastics.

In the end, Japan's Ramu Kawai claimed gold in the girls' event and Russian Sergei 'Bumblebee' Chernyshev won the boys'. Kawai, from Kawasaki City, started breaking when she was five – she trains outside her day job, from 10pm until 1am. Chernyshev is a former child gymnast and construction student from the central

Russian city of Voronezh. His coach is his father, also named Sergei, who was one of the first organisers of breakdancing in Voronezh in the 1990s.

But even the creation of that judging system drew a few raised eyebrows in the breaking community, which had long been a loose collective of like minds but is now hurriedly organising itself to handle the structures of Olympic qualifications.

Lowe Napalan, who is the president of the Australian Breaking Association, has been dancing, judging and teaching for two decades. 'I train a minimum ten hours a week and that's on top of a regular job – but that's just minimum and I've been training since I was a young kid,' he says. When it comes to judging, he says the key is to keep it simple. 'One aspect is the music. Are they in time with the music? Another aspect is the physical aspect – are they athletic, are they doing dynamic moments?

'The other aspect is creativity. Are they showing how they interpret the dance? It's very dependent on the judges themselves. It's not a normal sport criteria. It actually can be very susceptible to bias, depending on the judge's preference.'

So, what will a dancer have to do to take out the gold? 'It could be anything, really,' Napalan says. 'One of the No. 1 B-Boys at the moment is a very creative B-Boy, so he's beating people who are doing crazy flying tricks [where a dancer spins on their head, or hands]. You just don't know, it's unpredictable.'

Unlike in figure-skating, where competitors choreograph their moves meticulously, the music for breaking is not chosen by the performer. Instead, a house DJ is in charge and runs the same sort of consistent style through the competition – it is then up to the breaker to interpret and respond.

Hamilton from Chunky Move is fascinated to see how the mainstream responds to the sport and whether it will achieve the IOC's aim of getting more young people interested in the Olympics. 'Most people in the mainstream aren't necessarily aware of the impact breakdancing has had, and its structured competitions around the world that have been going for over 30 years,' he says. 'All you've got to do is do a search on Instagram and see where breaking is at already, in terms of its presence and its profile.'

Critics argue freestyle breakdancers can't be judged against one another, but Bobbito Garcia, another original member of the Rock Steady Crew, has said that couldn't be further from the truth – hip-hop has always been competitive. 'Think about it like boxing,' Garcia told sports website theundefeated.com. 'You're stepping into a ring. You're about to battle another warrior. The mental fortitude required, coupled with the athleticism – hell, yeah, breaking is a sport!'

25

HOW DO WE AGE?

When we glance in a mirror, we can be taken aback – how did we get to this age? Yet the process starts before we are born. What happens in our bodies over time? And how can we embrace our later years?

Sophie Aubrey

I t's a fact that many of us don't want to face: with every tick of the clock, every single one of us is ageing. It feels scary. But it needn't.

'Being human and living our lives is all about change, and that's what ageing is, it's change over time,' says Professor Julie Byles, a social gerontologist and researcher at Newcastle University.

Ageing is intrinsic to the living species on this planet. But how we grow old, and the factors that influence the process, are complex and unpredictable. They involve everything from the way we behave as individuals to the environment we grow up in and the inescapable truths coded in our genes as well as the sheer luck that befalls us.

How we grow old, and the factors
that influence the process,
are complex and unpredictable.

'Ageing is universal but not uniform: it's universal because it happens in all cells and all species but it's not uniform in that we don't all go through it in the same way,' Byles says.

Australia has one of the highest life expectancies in the world, ranking ninth among OECD countries behind Switzerland, Iceland, Italy, Norway, Japan, Sweden, Israel and Spain. An Australian born in 2019 can expect to live to about 83, some 34 years longer than people born in the 1880s (in Japan, the average age expectancy is just over 84). And the older population is growing: today, about one in seven of us is 65 and over. By 2057, it'll be almost one in four. As the World Health Organization says, with good health a longer life brings opportunities: to pursue new activities, a long-neglected passion or even a fresh career.

And we may feel more empowered to make those extra years as

fulfilling and meaningful as possible if we understand how ageing happens – that it's a lifelong process, not just some switch that gets flicked in your 60s, says Peter Lange, a University of Melbourne clinical associate professor in geriatrics. 'There is a lot of nihilism about ageing and a lot of people think that disease is inevitable, that they'll go into a nursing home or develop dementia. And that's not true. But by believing it's going to be the case, they end up failing to take action to prevent it,' he says.

So how do we age? Can we slow the process? What do we gain as we get older? And how can ageism be not just harmful to others but, in fact, self-sabotaging?

How do we typically age?

While most of us don't start feeling the effects until at least our 30s, ageing starts when we do. 'It's happening all your life. It starts even before you're born, from the very first cell division,' says Byles.

Our cognitive processes peak about the time we're 20. In fact, most of our body's systems are thought to peak when we are between 18 and 30, says Leon Flicker, a professor of geriatric medicine at the University of Western Australia. 'That seems to be when the ageing process . . . starts kicking in and you have a progressive decline.'

Ageing is not programmed, though. As we move through the world, we suffer little bits of damage – it could be from sunlight, bacteria, a sprained ankle, a shonky DNA copy, bad food – that the body then works to repair. 'It's happening throughout every second of our existence,' says Flicker. But over time our physiological reserves drop, so we're left accumulating damage that our body gradually can't keep up with fixing – and this can manifest in all sorts of ways.

The changes we experience as we age are neither linear nor consistent, and there is an extraordinary level of variability among older people. 'The thing about ageing is it's affecting every system of the body and different parts get impacted differently and no two 80-year-olds will be the same,' says Dr Kate Gregorevic, Royal Melbourne Hospital geriatrician and author of *Staying Alive: The Science of Living Healthier, Happier and Longer*.

There are certain hallmarks of ageing on a molecular and cellular level. For example, Gregorevic says, damage accumulates in our DNA. One way is that each time our cells divide, the little protective caps on the ends of our chromosomes, called telomeres – which are often likened to the plastic tips on shoelaces – gradually shorten, which affects our ability to copy DNA properly. When DNA is damaged, over time more cells can die or become cancerous. More also become senescent – they stop dividing, which on the one hand defends against cancer but on the other hand means they take up space without contributing, causing inflammation and overstimulating the immune system, which can't keep up with removing them. This is why senescent cells are sometimes called zombie cells.

Then there are the changes we often notice from our 40s, says Lange: skin loses elasticity; hair turns grey as pigment cells in our follicles slowly die; we become long-sighted and need reading glasses as the lenses of our eyes stiffen; in some people, hearing dulls.

Muscle loss, or sarcopenia, is another typical part of ageing. One study observed that muscle mass decreases between about 3 and 8 per cent each decade after age 30 and the rate of decline is higher after 60. 'We know that with age we can lose muscle strength,' Gregorevic says, 'particularly we lose fast-twitch muscle fibres – they're the ones you use when sprinting or when you catch your foot on the pavement and need to steady yourself fast.' At the same time, bone density drops while ligaments and joints stiffen, becoming more at risk of injury, and taking longer to heal.

Also common are weakening lungs: the amount of oxygen we take in with each breath slowly diminishes. Our cardiovascular system is impacted, with blood vessels and the heart muscle typically stiffening. Many people also eat less, possibly because dulling smell and taste reduce their enjoyment. And there are certain hormonal changes, with people exposed to a higher level of cortisol (the stress hormone) as they age.

The pile-up of all these changes gradually increases our vulnerability to disease or 'insults' – such as a seemingly minor fall leading to bone fracture – and eventually leads to frailty, which is a loss of physical reserve that affects almost everybody by their 90s, whether or not they have disease. 'If you're frail,' says Gregorevic, 'your body is already working so hard at the best of times just to get through daily life, so when you get a cold it takes all your energy.'

Does ageing necessarily involve disease?

Changes in the ageing process can be amplified or even triggered by disease or lifestyle. 'There's a lot of confusion between what is a disease which is associated with ageing and what is an implicit process of ageing which happens to everybody,' says Lange. A good example is our minds. Brain tissue gradually decreases with age and it's normal for our memory to shift: it gets slower and less efficient, so it can take longer to recall information. But forgetting entirely and suffering from significant impairment is not universal, it's a disease – a form of dementia.

We are living longer than ever, thanks to health advances. Most of those extra years are good years, too. In 2014–15, almost three-quarters of Australians aged 65 and over reported they had good or very good health, according to the Australian Institute of Health and Welfare. But, of course, disease does become more common in older people: in 2015, cancer and cardiovascular disease were the leading culprits, while dementia, type 2 diabetes, chronic lung disease and osteoarthritis were also prevalent. Many of these illnesses are also the biggest killers.

'Ageing and disease aren't synonymous, they're just associated,' Byles says.

Flicker explains that cancer becomes more common as we age, for example, because our body's surveillance system is not functioning as well so it is less likely to spot and destroy bad cells. Meanwhile, we are more at risk of diabetes, in part because our body becomes less efficient at converting glucose to energy and requires more insulin.

Still, while everyone's reserves decrease with age, we aren't all similarly susceptible to disease. Some of us, Lange says, have higher baseline defences. So, even though someone may have pathologies such as hypertension and mini-strokes, both of which are linked to dementia, they won't necessarily develop a form of the disease.

What difference does the life you've led make to ageing?

There's a cheeky saying in gerontology, says Byles: 'If you want to age well, pick your parents.' Being born with as few DNA errors as possible gives you a good headstart in life; aspects such as a good education, financial security and access to nutritious food in childhood add to your stocks. 'You can be already on an un-level playing field, depending on what your early life is like,' Byles says. 'If you get to older age and you're big and strong, with a healthy brain, good education, a strong immune system, they all go into your reserve ... you can maybe cope with having not as strong muscles or a decline in condition because you have all these other things that support you.'

A quarter of how we age is determined by genetics, according to the World Health Organization, and the rest comes down to lifestyle and socioeconomic factors. 'At any age, exercise and good nutrition is good for you but by starting younger you're giving yourself a chance of being well in older age,' Gregorevic says. More than a third (38 per cent) of the burden of disease in older Australians was preventable in 2015, according to the Australian Institute of Health and Welfare. Smoking, poor diet, being overweight or obese and high blood pressure were key contributors.

COVID lockdowns, limiting exercise and socialising, underlined how important lifestyle is in ageing; Lange says the months in lockdown in Victoria in 2020 alone had devastating impacts on physical and cognitive function for older people. 'That was a "multidomain insult" that produced really terrible effects on my patients,' he says. 'Most haven't recovered to their baseline level.'

But there are other factors outside of our control. Social disadvantage is a big one. A 2020 study of 5000 Britons found that lower socioeconomic status led to an accelerated decline in ageing. The researchers point out that the rich tend to have, for example, better access to parks and fitness centres as well as mentally stimulating activities (social clubs, the arts), which all help bolster physical and mental function. Meanwhile, those living in poverty usually experience more life stresses, which affect health, and they can also be exposed to more environmental pollution.

This disparity in ageing is reflected in Australian government policy. Subsidies for aged care services usually kick in at 65 but if you are either Aboriginal or Torres Strait Islander, or homeless – and so at higher risk of health issues and financial inequality – you are eligible at 50 and if you are both, eligibility drops to age 45. And the life expectancy of Indigenous Australians is about eight years less than for non-Indigenous Australians.

Aunty Geraldine Atkinson, a Bangerang woman and co-chair of the First Peoples' Assembly of Victoria, says more must be done to support healthy ageing in Aboriginal communities. 'Being brought up in poor housing, abject poverty, poor nutrition, a whole range of other things, all of that has contributed to our elders having chronic health conditions today,' she says.

▍ Can we delay ageing?

Much research is being done to find a magic anti-ageing pill but there is no strong evidence that any supplements or medications work so far. Scientists are attempting to find treatments that could lengthen telomeres, for example, or remove senescent cells via drugs or gene editing in the hope these could slow ageing.

The medical community is keeping a close eye on clinical trials underway to determine the effects of the drug Metformin, which is normally prescribed to manage blood sugar in diabetics but which has shown broader age-targeting potential. Research previously suggested diabetic people who took the drug outlived non-diabetics who didn't, and it has been found to delay ageing in mice, although in high quantities it was toxic. The American Federation for Aging Research is examining whether it can also prevent heart disease, cancer and dementia.

Still, any geriatrician will tell you a balanced diet and regular exercise are key to supporting healthy ageing and, in turn, a longer life. Stimulating your mind helps minimise disease risk. 'Just like our muscles, use it or lose it – our brains are like that as well,' Gregorevic says. 'One of the best things you can do for healthy ageing is just to keep having a go at things – not brain training but staying engaged in life, socialising, learning new skills, learning a language.'

It's never too early or too late to start. Byles encourages people to start planning for old age when young, as you would with superannuation. 'If you're not doing it by your 50s,' she says, 'that's when you must ask yourself, "Where do I go from here, what do I want to protect in terms of bodily, social and mental functions?" Make changes that will maintain your wellbeing – and keep you pushing your capacity to do things you enjoy.'

The aptly named Professor Norman Lazarus lays out why he considers exercise, eating well and mental health a 'trinity' in his book from 2020, *The Lazarus Strategy: How to Age Well and Wisely.*

The King's College London physician and researcher overhauled his lifestyle in his 50s and became a champion cyclist at 66, an age he says he'd expected to start having difficulty getting out of a chair or opening jars. At 85, he still cycles, trains at the gym, walks with his wife, watches his diet and works at the university.

Lazarus encourages people to first accept they are going to get old and then change their lifestyle – to truly enjoy the journey of ageing and retain their independence. 'At every age, I change my behaviour so that I can do the best I can with the physiological systems I have. And I don't look for immortality.' The trick is to prioritise things you love so you stick to them, he says. While he and his wife now walk on flatter, gentler trails instead of on multi-day hikes, for example, they still relish the time together.

He even uses the term 'exercise deficiency diseases' to emphasise how key physical activity is. A comprehensive study in 2015 found 26 common illnesses could be positively affected by exercise, including cardiovascular disease, type 2 diabetes, obesity, dementia, osteoarthritis, osteoporosis and some cancers. 'The most effective anti-ageing option we have is exercise,' Lange says. 'It produces beneficial effects for pretty much everything we've ever looked at.'

Exercise improves muscle strength, balance, bone density, and the immune, cardiovascular and respiratory systems. It boosts mood and supports brain and spinal health, too. Getting out and being active with people is important for cognitive stimulation; and, by stressing the body, you're also getting it used to dealing with small perturbations.

Lange says just a small amount of activity can make a meaningful difference, and resistance training is particularly important – he has seen this transform once-bedbound patients in their 80s and 90s. He encourages people to introduce incidental movement into their day: carry your shopping bags instead of wheeling a trolley, walk instead of drive or tend to your garden.

Even if you have the healthiest exercise, diet and social regimen in the world, though, you could still suffer from disease when older (or younger, for that matter). 'The harsh reality is that no matter what you do, you're going to die. And you can live the perfect lifestyle and still get cancer. Nothing is certain,' Gregorevic says.

> *The harsh reality is that no matter what you do, you're going to die.*

It's why we must be careful not to judge people for the condition they're in when they're older. Byles points out that people with certain illnesses, such as diabetes or lung disease, often get blamed. 'Some of it is preventable but not all of it. Some of it is by virtue of the fact you've been around a long time and have had a lot of chances to accumulate a problem,' she says.

How does ageism affect things?

It's a familiar feeling, being caught by surprise by your own age: it might be catching a glimpse of your face in the mirror and realising you don't quite recognise yourself, or going to run after a grandchild and realising your muscles don't mobilise like before.

Part of the reason you don't notice your own ageing is because the changes are far too subtle, Flicker says. And people don't actually think they're changing with age – our vision of 'self' is deeply ingrained. Byles says: 'I have a theory that everyone thinks they're 30. You have a concept of yourself, so we always think we're younger, which is great, but then we can get a shock.' She believes part of this is our own ageism, where we value our youth as more relevant.

Indeed, we live in a culture that glorifies youth. In Australia's 2021 Royal Commission into Aged Care Quality and Safety, commissioner Lynelle Briggs found that ageism is systemic in Australia. 'The acceptance of poorer service provision in aged care reflects an undervaluing of the worth of older people, assumptions and stereotypes about older people and their capabilities, and ageism towards them. This must change,' she wrote. And a WHO report concludes that ageism is widespread with far-reaching consequences – its global survey of more than 83,000 people found that one in two had ageist attitudes.

One of the great tragedies of ageism is that people internalise it and develop a negative bias against their older selves, Gregorevic says. On the flipside, a study has found that people with a positive attitude about ageing are less likely to develop dementia. One common example of self-sabotage is that while young exercisers crave feeling out of breath, many older people, doubting their capabilities, take feeling out of breath to mean they should avoid being active, Lange says. '[But] that's exercise and it's getting you used to those challenging activities, and the next time you get sick and need some extra heart and lung function, it'll be there.'

That older people are incapable of using technology is another false preconception, while the retirement age generates another myth, says Professor Linda Rosenman, board member of the Australian Association of Gerontology. 'I think it's really important not to categorise everybody older than 65 as old,' Rosenman says. 'This is just the age that people become eligible for government pensions. There's nothing magical about this age ... It doesn't mean people are all of a sudden decrepit.'

Not all Australian communities suffer from ageism: 'It seems not right, foreign, kind of,' says Aunty Geraldine, explaining that Indigenous communities focus not on what's lost with age but instead on the wisdom that's gained – something she thinks the rest of the country could learn from. In Aboriginal culture, an elder is someone who is recognised for their knowledge and ongoing contribution to their community, and they are not neglected. They are traditionally referred to as 'Aunty' or 'Uncle'. 'We always respect and value our elders, respect their cultural authority, respect the stories they told us ... and that gets passed down,' Aunty Geraldine says. 'You become an Aunty not just to your bloodline, but to other younger people as well.'

> *Indigenous communities focus not on what's lost with age but instead on the wisdom that's gained.*

It's different in other countries, too. 'When you're in Beijing and you go for a walk to parks, they're full of older people doing taichi, dancing. There's no sense that, "I'm too old to do that." If we go to parks, we see young people playing football,' Byles says.

Japan is known for its super-ageing society. It not only has the highest life expectancy, according to some statistics, but the world's highest proportion of older citizens, with seniors aged 65 and over accounting for more than 28 per cent of the population. The nation has introduced measures to support healthy ageing, including the way cities are designed (for example, some pedestrian crossings have a second button to give extra time). A high proportion of older Japanese continue to be engaged in community activities, live in their own home and have strong cognitive and physical abilities.

Byles, who helped create the anti-ageism Every Age Counts campaign, says people could be enjoying life more as they age. 'Ageing is a great individual and societal success. We should be embracing it.'

What do you gain as you age?

Actress Jane Fonda was onto something when she said, 'As I started getting older, I realised, "I'm so happy!" I didn't expect this! I wasn't happy when I was young.' The fact is we do typically become happier as we age, with research showing older people tend to have brighter moods and fewer symptoms of depression and anxiety than younger counterparts.

Byles says it's possible this is because older people largely do things that bring them satisfaction. 'You can't do all the things you used to do . . . so you are actually focusing on things that are more important to you.' Plus, she says, we often care less about what other people think, which can be really liberating. And there is a sense of contentedness that comes from understanding, with time, where you and the puzzle pieces of your life fit.

There are also certain remarkable improvements in cognition. People keep improving their vocabulary well into their 60s and 70s. And while brain speed and working memory peak in early life, this doesn't make 20-year-olds equipped mentally to run the country. Older people have what's termed 'crystalline intelligence', Flicker says. 'You use your brain to solve problems throughout your life . . . [and] each time you learn a new strategy you can apply it to other problems, and that's wisdom.' This likely also contributes to happiness, Lange says, because experience tells you that even when things seem disastrous, you'll get through to the other side.

Reflecting on her own life, Rosenman considers herself to have become more patient and tolerant, and she enjoys having more time. 'You're not trying to climb the greasy pole yourself any more and you're much readier to mentor and help other people,' she says. 'In many ways, life is a lot more enjoyable than when you were racing off to work and herding the kids . . . [and] grandchildren are a big bonus.'

How do you deal with the fear of nearing the end of your life?

There is, of course, an existential aspect to ageing. 'That can be quite difficult,' Lange says. 'I've met patients who have lived too long and have outlived their partners and friends and even sometimes their own children, and they have increasing physical disability and sensory limitations.' But, he says, most people reach a point where they no longer fear dying. 'It's quite unusual to have ongoing concerns about the prospect of death. People usually accept it will come . . . which is quite freeing.'

Lange hopes people focus on ageing well rather than fixating on the idea that getting older means they're approaching death – which is only one moment at the very end of a life journey. Flicker agrees: 'I think people eventually realise that if you're not dead yet then maybe life is to be lived, and . . . really you should try to enjoy the living as much as you can, no matter what is happening at the time.'

And it's up to all of us to talk to each other to help reframe the way we view our older years. As the aged care royal commission report outlines, 'There are everyday things that all of us can do to enable older people to live their lives to the fullest extent possible . . . Older people should also be encouraged to think about what it is that would make them happy, and to have some goals or objectives for each day or week that give purpose.'

It's up to all of us to talk to each other to help reframe the way we view our older years.

While many people worry they will end up in a nursing home, the reality is most Australians do not. The same report says about 80 per cent of Australians use an aged care program at some stage before their death but for most this means at-home support.

'Nobody is saying the last years can't be difficult. You can be faced with losing people you love while also dealing with your own debilitating health problems, but the vast majority of people will be healthy enough to live independently for most of their life,' Gregorevic says. 'We need to remember that life is finite and, in a way, that that's a real gift.'

Like all of us, Lazarus feels taken aback when he sees his older face in the mirror but says there isn't any age that he wishes to be frozen in. 'I'm not sure where I put myself. I see this old man and I see myself, but . . . do I wish I were 20 again? Not really,' he says. 'I can't place myself in any decade because I've really enjoyed myself.'

26

HOW DID BITCOIN GO MAINSTREAM?

Once a nerd's hobby, Bitcoin is huge business.
What does a story about pizzas have to
do with it, and is it true that if you forget
your password, you lose your dough?

Dominic Powell

In May 2010, on a sunny day in Florida, computer programmer Laszlo Hanyecz bought two pizzas online for the equivalent of US$30. If he was to have performed the same transaction in, say, mid-2021, the meal would have cost him nearly US$600 million. That's because Hanyecz paid for his dinner in Bitcoin – 10,000 of them. Hanyecz's pizza purchase is now a part of the Bitcoin story, widely regarded as the first commercial transaction using the mysterious cryptocurrency and one that helped catapult it into the mainstream.

In 2010, the only places you might have heard of Bitcoin or other cryptocurrencies would have been in the dark depths of an internet message board used by nerdy teens looking for discreet ways to purchase drugs (or pizza). The digital currency's roots are closely linked with anti-establishment or libertarian movements, and early adopters touted its independence from banking institutions and freedom from government oversight as major benefits.

Today, Bitcoin and other cryptocurrencies are storming the world, offering a new asset class and catching the eye of investors everywhere, including one of the world's richest men, entrepreneur Elon Musk. Every day Bitcoin changes hands hundreds of thousands of times all over the globe. It is often referred to as 'digital gold'. But it is also extremely volatile and entirely unregulated, with no help desk if you get into trouble.

So, what's the attraction of Bitcoin? And how does it work?

What is Bitcoin?

In 2008 – two years before the storied pizza delivery – an anonymous person or group of people known only as Satoshi Nakamoto released an explanatory paper, 'Bitcoin: A Peer-to-Peer Electronic Cash System'. It set out Nakamoto's idea for an electronic version of cash that would be secure, stable, trusted and trackable, all without the need for a central financial institution.

In essence, Nakamoto's original vision for Bitcoin still holds true. It is a cryptocurrency: a digital currency that's encrypted, which means it's programmed to make it incredibly difficult to counterfeit.

You won't find the local Bitcoin headquarters in a CBD, nor will you hear about the exploits of Bitcoin's chief executive. You won't be able to put any shiny Bitcoins in your back pocket, nor call up the Bitcoin hotline when you've got an issue with your Bitcoin credit card. None of these things exists. There's not even a central server nor system that runs Bitcoin. Instead, the network exists on a web of millions of computers across the globe, each linked to one another and tasked with verifying, cross-referencing and processing transactions on the network.

Huh?

Let's say you decide to send your mum a Bitcoin – or a fraction of a Bitcoin, which is all that most people could afford these days – as a birthday present. You and your mum would each need to have a Bitcoin wallet (which is like an account); you'd need to find an online exchange (there are plenty); you'd set up an account linked to your bank account; then buy the Bitcoins and send them to her wallet.

When you hit send, thousands of computers, called nodes, would all independently check that your transaction was above board – that you were good for it. If a majority agreed, your transaction would be chronologically added to a long public list – or chain – of every Bitcoin transaction in existence and the sale would go through.

Where does 'blockchain' technology fit in?

Every ten minutes, the computers managing the network package all the transactions received during that time as a 'block', which is linked to the preceding block.

Why ten minutes? That's just the time coded into the design when this system was first made. The point is, these blocks are unable to be modified or changed and can be traced all the way back to January 2009 when the network was first switched on.

It is these blocks that form the 'blockchain' technology upon which almost all cryptocurrencies are based. There are now hundreds of cryptos similar to Bitcoin. There are even cryptos made as jokes (albeit that have value) such as Dogecoin, based on an internet meme about a sheepish-looking Japanese dog and created as a kind of satirical jibe at cryptos.

The reasons for running currencies using blockchain technology are numerous. For example, if one computer processing a transaction crashes, millions of others can step in and pick up the slack. Similarly, if someone was to try to dupe the system with a fraudulent transaction, every other node could reference it with its own copy of the blockchain, see that it was invalid and refuse to verify it.

Blockchain technology also creates a transparent ledger visible to anyone who cares to look, on which nothing can be changed, modified or hacked. The transaction itself is recorded but it doesn't show the identities of people or companies involved, so anonymity is ensured, which can be good or bad, depending on how you look at it. The transparency and immutability is attractive to many Bitcoin fanatics, and to a growing number of investors generally, who decry what they see as the murky operations of many financial institutions. Bitcoin, remember, was rolled out in 2008, in the midst of the global financial crisis.

Blockchain technology has other uses, too – it's not all about cryptos. For example, it has been used to develop 'digital collectibles' known as non-fungible tokens, or NFTs. These assets, which tend to take the form of a piece of art or a short video clip, are essentially a computer link that their owner can click on to look at them, placed within the blockchain. They're like any collectible in the sense that they're limited-edition and able to be transacted among people. The owner of an NFT has their exclusive ownership registered on the blockchain, unable to be altered or changed even if the NFT can be copied. It attracts the kudos of owning an original work of art. Some NFTs fetched millions of dollars at auctions in 2021.

And many companies make products using blockchain that have nothing to do with cryptocurrency or NFTs. For example, Sydney-based startup Lumachain is using blockchain to tackle transparency in global supply chains because the unalterable ledger is a record of an item's provenance.

What does mining have to do with Bitcoin?

Making money is, of course, the driver of the Bitcoin boom – and not just for individual punters at home. Every day there are an estimated 400,000 Bitcoin transactions – anything from people moving their own Bitcoin between wallets, as you might do with bank accounts, or buying and selling Bitcoin, or using Bitcoin to buy other cryptocurrencies. The process of verifying and recording these moves requires incredible computing power, and network operators can hardly be expected to keep the computers running on pure altruism. Instead, these operators get rich from 'mining' Bitcoin. Mining is the crux of how the self-generating Bitcoin sausage is made.

Every ten minutes, the first computer (or group of computers) to announce to the rest of the network that it has successfully verified all the transactions in a block is rewarded with 6.25 Bitcoin, worth just over half a million Australian dollars (the value goes up and down). This is called a block reward, from which miners make a tidy profit – by selling it into the wider Bitcoin market through a cryptocurrency exchange, which acts like a highly decentralised mini stock market.

Mining could once be done on any old laptop. Back in 2010, those same nerds in internet chatrooms could have landed Bitcoins in mere seconds. There was actually a website called the Bitcoin Tap on which users could receive five Bitcoins just by entering their wallet (account) details. But another crucial ingredient in the Bitcoin sausage is this: reserves are finite. Bitcoin was programmed in such a way that there will only ever be 21 million coins in existence. With about 18.6 million coins already in circulation, the computational difficulty of verifying transactions will continue to get exponentially harder – so hard that the rate of release will slow and Bitcoin's supply won't be depleted for another 120 years. This is why mining Bitcoin requires serious processing power and a bit of luck, the luck coming from whoever gets over the line first to verify a block and seize their 'Eureka!' moment.

It's no surprise, then, that Bitcoin mining has become a business in itself. Massive companies, from Reykjavik to Siberia to Amsterdam, are tasked with overseeing warehouses full of computers dedicated entirely to mining Bitcoin. Around the world, miners pocket tens of millions of dollars a day in block rewards. Miners also pocket fees paid by Bitcoin users for each transaction that occurs within a block. If you simply wanted to transfer your Bitcoin between two wallets (accounts), in May 2021 that would have cost you around AUD$20, although the ever-fluctuating price of Bitcoin means this has been as high as AUD$80.

Running these computers is also a power-hungry process – we're talking about warehouses stacked full of processors plugged into the grid. Unlike other warehouses full of internet servers, these Bitcoin miners aren't providing a great deal of public utility. And yet the Bitcoin Power Index, run by news site Digiconomist, estimates the Bitcoin network has an annual carbon footprint comparable to that of Peru (population 32 million). The electricity consumed in just one Bitcoin transaction – with all those computers crunching tough equations – could power an average US household for more than 23 days. This is why many sustainability advocates prefer cryptocurrencies such as Ethereum, which have far cheaper running costs and don't use as much energy. It's likely that as long as Bitcoin and other cryptocurrencies require immense amounts of processing power to operate, the green credentials of cryptocurrencies will continue to be an issue.

How did Bitcoin catch on?

Bitcoin's value can be a tricky thing to understand. Why should something with no product or commodity tied to its value, and that generates no cash flow of its own, be worth, say, $60,000 apiece? Isn't it all just ones and zeroes floating around on the internet?

Well, yes. These are valid points and ones often raised by Bitcoin's detractors, who have frequently denounced the asset as a scam comparable to a Ponzi scheme, believing the coins maintain value only as long as there's a steady stream of greater fools willing to buy in.

But others call Bitcoin 'digital gold' insofar as both gold and Bitcoin are finite – you can't simply create more of them – and they take considerable effort to extract. And the economic context factors into its appeal. Bitcoin became increasingly attractive for investors in 2020 as central banks around the world pumped their economies, printing money at a rapid clip in response to the COVID-19 pandemic, sending interest rates to record lows. Having money in the bank has been generating measly returns, meaning investors have been looking further afield for assets that may appreciate faster. Major international funds such as BlackRock have begun to invest in Bitcoin, alongside Wall Street legends such as Stanley Druckenmiller and Paul Tudor Jones. And for good reason: the price of Bitcoin over the first half of 2021 rose 58 per cent, far greater than the ASX200's 5.6 per cent gain.

Then there's the influence of fame. In February 2021, Elon Musk announced that his automotive company, Tesla, had invested US$1.5 billion in Bitcoin and would begin to accept the cryptocurrency as a payment option for its electric cars. Following the news, the price of Bitcoin spiked nearly 20 per cent to a record high of US$72,800 as investors flocked to emulate Musk's surprise purchase. However, the price later fell more than 20 per cent as Musk walked back his decision that Tesla would accept the currency as payment.

Still, Musk has long been an advocate for cryptocurrencies and many prospective Tesla owners are young and male, intersecting with the core demographic of Bitcoin investors. The company also signalled it could look to invest further in cryptocurrencies, noting it may 'acquire and hold digital assets from time to time or long-term'.

And there is a social factor at work. Bitcoin's price is in part driven by its devout, almost religious, followers who extol the currency's freedom from governments and banks and believe it will be a leading global currency in years to come.

Asher Tan was working as an economist on Melbourne's Collins Street in 2011 when he first caught wind of Bitcoin. He recalls reading distinguished US economist Paul Krugman's haranguing of the then-obscure currency. 'I read his stuff on it enough times and thought, maybe he's my idol but maybe he's wrong.'

Tan now runs one of Australia's most prominent crypto exchanges, CoinJar, but back in 2011 he says the idea of a social movement driven by the internet and the democratisation of finance was quite novel – and polarising. 'At its core, it's a message of old versus new, bottom-up versus top-down, and Bitcoin is the medium through which a lot of people choose to express this,' he says. 'Bitcoin means different things to different people but the most important thing is that it's still here now. Whatever your take, it's still meaningful and relevant.'

Are you a Bitcoin type of investor?

Remember those pizzas in Florida? In Hanyecz's day, buying Bitcoin required proficiency in the dark arts of the internet, as exchanges were often difficult to access. Today, exchanges have become far easier to find, requiring a mere Google search and a brief sign-up process followed by an identity check. Or you can do it all on your phone: apps such as Coinbase are popular for buying small amounts of Bitcoin or other cryptocurrencies.

But, while it might be easy to make a purchase, be warned: the potential for screwing up a Bitcoin transaction is quite high. If you're someone who forgets passwords easily, for example, it might be best to stick to more traditional investment choices. Here's why.

After buying Bitcoin (or another cryptocurrency) from an exchange, the coins are usually stored and managed by the platform itself. This acts much like a trading account with a broker, with your exchange account linked to your bank account to make it easy to buy and sell – cash in your Bitcoin and the money simply drops into your account.

For small amounts of money, keeping your money locked up on an exchange is generally pretty safe. But if your purchases start to stretch into the tens of thousands, it might be time to buy your own personal crypto wallet. These devices look closer to a USB stick than your parent's leather bi-fold and connect via software on your computer, which allows you to transfer your coins across. Wallets can store any number of different cryptocurrencies and are a key part of trading crypto. As of May 2021, there were around 64 million active Bitcoin wallets. Only you can access your wallet.

When setting up a personal wallet for the first time, you are presented with two crucial pieces of information. Firstly, each wallet has a 'public key', which is a string of numbers and letters that allows you to receive coins into your wallet, much like a BSB number and bank account. If your mum wanted to send back that birthday gift you gave her, she'd need this number.

Secondly – and far more importantly – each wallet includes a private key, which is a secret number that grants full access to your stored coins. These often come in the form of a 12- or 24-word recovery phrase, comprising a string of random words that translate into your private key. This phrase is the master key to your Bitcoin and should be protected with your life. It's important to note your Bitcoins are not actually stored in these wallets. The key gives you only the right to access your Bitcoin, which is stored on the blockchain. Wallets just serve to store and protect your private key – if you lose your wallet, your private key will still allow you to access your coins so long as you've noted it somewhere. If you lose your wallet and you lose your private key, your coins will be lost forever. Around AUD\$190 billion in Bitcoin is currently lost or inaccessible.

The cryptocurrency world also remains almost unregulated for now. The lack of regulation goes to the heart of what many Bitcoin fanatics enjoy about the currency, with it being largely outside of government control, but this also means an exchange you use to buy and sell crypto could disappear overnight – with your funds. In Australia, the government will usually have your back if a bank or financial institution you invest with collapses, but there are no such protections in the world of crypto.

Cryptocurrencies are immensely volatile. Crypto markets are also often dominated by 'whales': investors with massive amounts of cryptocurrency who have the ability to move markets on a whim. If they sell even a small portion of their holdings, they have the potential to send prices crashing.

Investing in Bitcoin is not for the faint-hearted. Expect a wild ride. They are 24-hour markets, which makes them impossible for traders (who need some sleep) to monitor constantly. Some investors even set alarms to be notified of major price swings in the middle of the night. In fact, in the time it takes to make a cup of tea, Bitcoin's price could fluctuate by \$1000. So don't be surprised if that \$60,000 Bitcoin you just bought is worth \$30,000 the day after. Just be thankful it's not two warm pizzas.

WHY ARE NATIONS LAUNCHING CRYPTOCURRENCIES?

The rise of cryptocurrencies has also prompted a number of countries to explore, and even launch, their own form of digital currency. China, Singapore, Russia and Sweden are among those with government-backed digital currencies in the works.

Known as central bank digital currencies (CBDCs), these are effectively just a digital version of cash. They're backed and issued by central banks, can be used as a mode of payment, and are serialised and trackable just like any bank note. Due to this, CBDCs are only rudimentarily similar to cryptocurrencies. While both can be used as a means of transferring money, cryptos such as Bitcoin are decentralised and untraceable by design. Indeed, most central banks developing CBDCs have stressed that their currencies are not crypto-assets.

Interest in CBDCs has skyrocketed in tandem with interest in cryptocurrencies. Governments are starting to wake up to the benefits of digital currencies. Crypto-purists would also tell you that financial institutions are growing wary of the power of cryptocurrencies such as Bitcoin, and are looking to introduce CBDCs in order to establish a presence in that space. Australia's Reserve Bank has flagged an interest in launching a digital currency, and has partnered with Commonwealth Bank, National Australia Bank, Perpetual and blockchain company ConsenSys to investigate its potential for Australia.

27

WHAT ARE CATS AND DOGS THINKING?

How does your dog know when you are sad?
Is your cat plotting world domination? How do
dogs and cats sense the world – and will they
ever be able to tell you about it in words?

Sherryn Groch

When my father was young, a dog saved his life by leaping on a striking snake. Decades later, I watched our overfed ginger cat 'mourn' Dad's death, lying every day in the garage where he once worked, rubbing his whiskers against Dad's tools.

No one who has lived with a pet cat or dog could deny they have feelings: affection, irritation, fear, perhaps even shame and jealousy, perhaps love. But is their behaviour always what it seems? Did my father's dog, who was rather unfairly named Bimbo, pounce on the snake to protect him or was some wolfish hunting instinct surfacing at last? Did Max the cat really wonder where Dad had gone or had he just decided to annex the newly available garage?

Dogs and cats have been our companions for thousands of years and remain fiercely popular (almost half of Australia's households now have a dog, for example). Yet serious scientific questions about their inner lives have been asked only in recent decades. Why is it that a dog always seems to know when you're sad? How did these two species evolve from the wild into our homes? Could they one day learn to talk to us? And are cats secretly plotting world domination?

How did cats and dogs become cats and dogs?

A prehistoric puppy thawed from Siberia's icy permafrost in 2018 could solve the mystery of how wolves first became man's best friend. 'Dogor' is remarkably well preserved, for a two-month-old who died 18,000 years ago, with teeth, fur, even whiskers still intact. And yet, despite extensive DNA testing to determine his species, it remains unclear so far.

'We don't [know] yet whether Dogor is a dog or a wolf or a bit of both,' says researcher David Stanton. Could Dogor be the missing link between wolf and dog? Most researchers agree pooches evolved from wolves between 15,000 and 32,000 years ago.

Dogs are the wolves that came in to sit by the campfire, that learnt to work with our ancestors for food, helping them hunt and manage other animals, offering protection and now, increasingly in the modern world, companionship. Having evolved at our side, they can read our facial expressions. Even their patented 'puppy dog eyes' offer an evolutionary advantage – for dogs, life has become not so much the survival of the fittest but of the cutest.

'We're wired to respond to them, too,' says Melissa Starling, who both trains and researches dogs at the University of Sydney and has her own lively brood at home. 'We have no defence against puppy dog eyes.'

Still, dogs today remain 99 per cent wolf (even pugs). The tiny changes in their DNA affect the digestive system and the brain as well as the rate of their physical development (which helps explain the big differences between breeds). Essentially, dogs are well-mannered wolves that can eat grains.

Unlike dogs, cats are not social by nature, coming from a long, proud line of solitary felines including tigers and leopards. How then have we tamed these wildcats into wearing collars and posing for Instagram photos?

It seems it was largely their idea.

When agriculture began some 10,000 years ago, the grain attracted rats and mice, which eventually attracted smaller wildcats. Our ancestors quite liked this pest control service and so left out treats to encourage the cats to stay. DNA evidence now shows that domestication has helped cats take over the world – shipped out from their original home in the Middle East to just about every continent on Earth.

Yet cats have not been selectively bred to the same degree as dogs. Even today, they are still considered only semi-domesticated – the tiger in your living room. Within a few generations in the wild, house cats can revert back to larger, fiercer animals, the feral cats that now wreak such a toll on our wildlife. (This might account for all those mysterious sightings of 'panthers' and 'leopards' in remote corners of Australia.)

Most scientists agree that animals can have complex, even profound emotions.

Do cats and dogs love us?

Any wild animal might become tamer if handled by humans from a young age. And, as our urban sprawl keeps on sprawling, there are even early signs that other animals such as coyotes might be growing more familiar with us. But cats and dogs are different. While they still need to be around humans early to grow up 'normal' (that is, well-behaved), it's not just nurture at play. Thousands of years by our side has created some serious chemistry, too.

Both species have been recorded getting a boost in oxytocin – the hormone released when we're in love or bonding – while being patted by their owners. Dogs seem to get a bigger hit of this 'molecule d'amour' than cats. Indeed, both dogs and their owners can produce it just by staring into each other's eyes (this does not happen between wolves and humans). When US neuroscientist Gregory Berns began training dogs to lie still in noisy MRI machines, he found another promising sign of affection: the reward centre in a dog's brain lights up more powerfully in the presence of its owner's scent compared to when it's sniffing other humans or dogs.

Twenty-five years ago, Jeffrey Masson, a former psychoanalyst from California, wrote his first bestselling book on animals, *When Elephants Weep*, and found himself accused of that great intellectual sin: anthropomorphising (attributing human traits to animals or objects). But, today, most scientists agree that animals can have complex, even profound emotions.

Masson himself believes dogs are now better at loving than people are. 'Just as you can't be as content as a cat, no one will ever love you like your dog,' he says. 'They love purely.'

He recalls a tiny puppy he rescued from a car crash when he was studying in India. The pair became inseparable until it was time for Masson to return home to the States. He found the dog, which he'd named Puppy, a loving new family and had the tearful goodbye. But the next day as he was farewelling a professor at his university, there came a sudden banging and scratching at the door. 'It was Puppy. I still can't understand it. Someone even swore later they had seen Puppy hop on and off a bus to the university.'

With cats, love is a little more complicated. They are famously independent, even aloof. They require wooing, chin scratches, multiple openings of tin cans. And even then, it's hard to ignore that cold look of disdain from the top of the bookshelf while doing your morning yoga routine.

'Cats are so graceful, everything they do is aesthetically pleasing,' says Masson. 'Of course, not everything we do is pleasing to them ... But they choose us, they're capable of deep affection. If you've ever shared a bed with a cat, you'll know. They really settle in, they purr.'

Still, cats are not necessarily faithful. One of Masson's own once migrated next door. 'And my neighbour didn't even like cats.'

The fickleness of cats is why many scientists are still loath to study them in the lab. To test one cat, you need three, they will say, as the other two will probably withdraw consent halfway through. 'They are notoriously difficult,' says Federico Rossano, who works with animals of all shapes and sizes as director of the Comparative Cognition Lab at the University of California. 'We saw a huge boost in research into dogs starting from the 90s but we haven't seen the same with cats. But when they do participate, they can give us great results, even match the dogs sometimes.'

At Oregon State University, Kristyn Vitale and Monique Udell have run groundbreaking experiments showing that cats display the same signs of attachment to their owners as dogs. Even more astounding, they've shown that cats prefer interacting with people over toys and, yes, food, and will seek out humans who pay attention to them.

Because domestication has made cats and dogs reliant on us, both species live in a kind of permanent juvenile mindset, Vitale says, where we become almost a surrogate mother. It's why you see cats purring and 'kneading' with their paws – the same behaviours kittens exhibit towards their mothers.

How do cats and dogs perceive the world?

Humans tend to *see* the world first – dogs smell it. For dogs, smell is the primary sense – and the world is one aromatic buffet of informative scents. The nose of a dog is at least 10,000 times more powerful than your own. It can sniff out storms before a whisper of rain is on the air, find cancer cells in our blood, or catch a familiar scent up to 20 kilometres away. We have bred dogs to help us hunt and now we train them for more modern jobs such as sniffing out bombs, drugs, even COVID. Squads of coronavirus-sniffing canines, for example, have been trialled in airports around the world (but don't expect a pooch to replace the eye-watering Q-tip test any time soon).

> *For dogs, smell is the primary sense – and the world is one aromatic buffet of informative scents.*

'Smell is such a minor sense for us it's hard to appreciate what the world is like for them,' Starling says. When a dog stops on a walk, it's a little like checking social media; they will often sniff around to see who else has been there and leave behind a 'post' of their own.

But, as it turns out, cats may make even better sniffer animals than dogs (if they could be persuaded to take up the job). Research by Vitale and others has found that, while cats have fewer smell receptors than dogs overall, they have a more refined palate, with more of a particular protein in the nose believed to help animals differentiate between smells. Vitale, who once ran her own version of puppy preschool for kittens, insists cats can be trained, too.

'People think it's bananas but they are quite capable of learning,' Starling agrees. 'Especially if they're food motivated.'

I don't have a dog at the moment. But, in the interest of research (or perhaps as a cry for help), I have started taking my two cats for walks – on leads. Masson assures me I haven't gone mad. He confesses he used to take his *six* cats for walks to the beach when he lived on the coast of New Zealand. They'd stroll down in the dead of night when no one else was around. The family dog would come, too, and the chickens and rabbit would sometimes follow. 'It was some of the happiest times of my life,' Masson says. 'The cats would race off and then hide and leap out and scare the dog. They loved it!'

To my surprise, so do mine. They meow at the back door. They drag out their harnesses. They throw their little heads back in the breeze and flop in the grass, eyes closed, purring. They even let me lead them, sort of, if I ask really nicely, if we don't go too far. Their ears are constantly moving, triangulating sounds.

A cat's hearing is even better than a dog's (perfect for catching every stealthy opening of the pantry door) and both animals can pick up much higher frequencies than humans. Where we have the advantage is in catching minor differences in pitch, say for appreciating a Vivaldi concerto. Humans are unusually good at this in the animal kingdom, rivalled only by bats. But, as biologist John Bradshaw writes in *Cat Sense*, a cat could never be trained to sing in tune, which is 'bad news for Andrew Lloyd Webber'.

The vision of cats and dogs is also less precise than humans because it's designed to help them hunt, catching movement more than detail. So, while they have a wider field of vision (240 degrees for dogs and 200 for cats compared to 180 for humans), cats in particular struggle to focus close up. They also don't see in the rich spectrum of colour that we enjoy, but neither are they completely colour blind; the world is painted in mostly blues and yellows for cats and dogs. And they can see much better than we can at night thanks to a reflective layer behind the eye that helps catch even the faintest glimmer of light (and gives them that distinctive green eyeshine). Some scientists even think they can see in ultraviolet.

Starling watches one of her dogs, Kestral, a Portuguese Podengo bred originally to chase rabbits, shoot through a tiny hole in the fence without slowing a beat. 'She's so fast, she'll go straight for a gap I don't even know is there, like she's got a map in her head.' Of course, Kestral returns. Dogs like to stay near their humans, while cats tend to stick to a more vaguely defined home territory (that sometimes extends into the neighbour's kitchen). Armed with a famously gravity-defying sense of balance, as well as sensitive whiskers to judge space, they will scale fences, trees, roofs and squeeze through every nook and cranny.

Yet, while cats often fight other felines who cross into this range, research (and spy cameras) have revealed they are not as militant about defending it as once believed. A cat investigating another's scent in their yard, or on their owner, might be more curious than jealous, Vitale says. When they rub on us, she thinks it's more to say 'this is someone I know and like' rather than 'this is my human, not yours'. Stray cats sometimes greet each other in the same way, and even have the unusual habit of forming colonies that loosely resemble lion prides.

Can pets teach us things and sense our feelings?

It had been a sweet but unlucky puppy named Harvey who prepared me for the suddenness of death, the chest-splitting pain. I was eight when he died. For many children, Masson says, the death of a beloved pet will be their first experience of grief. 'The good thing is people don't really say "It's just a dog" any more. Pets help teach children empathy, too.'

And they can do more than that. At the end of his mother's life, Masson would take his dog Benjy into her nursing home and watch him lift the shadow of dementia from her eyes. As a reporter, I've seen it, too, the therapeutic power of an animal, not just in aged care but in schools among children with a history of trauma. They'd tell me the moment they reached out their hands to touch the therapy dog, they instantly calmed down, as if the dog was a kind of talisman.

'Dogs want to be near us,' Starling says. But, while she agrees most pooches know when the humans around them are sad, and many will try to fix it, she notes sometimes, in coming over, they are really seeking comfort themselves.

'They're saying, "This is stressing me out. I need some pats." They're like children in a lot of ways. Some dogs make excellent therapy dogs, just like some people make great counsellors. Some can't handle it.'

Starling, whose own dog Erik the Tall is being treated for clinical anxiety, calls canines 'social acrobats'. 'They can handle a lot of situations and live in harmony with lots of other species [even cats]. But they put up with things humans do, too, even when their body language tells me they don't really like it.'

A wagging tail, for example, is not always the sign of a happy dog – if it's high and stiff, the dog could be agitated. 'That's when I look at the face,' Starling says. 'If there's tension there, if they're making jerky glances around. And then, if they suddenly close their mouth, that's a problem. Often, what happens next is quite dramatic.'

Likewise felines, and their famously inscrutable 'resting cat face'; it can be hard to know when a cat is about to scratch or purr. Even Grumpy Cat, made famous by the internet for her signature frown, was not really grumpy, just misunderstood (and with a rare medical condition). Vitale has four cats at home and says felines are easy enough to read if you know what you're looking for. Slow, sleepy blinks can actually function like a smile. But, when the ears go back, and their tail is swishing with sudden speed, it's probably time to take your hand back from that tummy rub. And a cat 'chattering' at a bird could be a sign of frustrated hunting instinct – and a need for more play.

> *Felines are easy enough to read*
> *if you know what you're looking for.*

But what if our pets could just tell us what they wanted?

Primates might be closer to humans in DNA (and brainpower) but researchers say cats and dogs are much better at *understanding* us, and that means they might also be better at communicating. In 2018, speech pathologist Christina Hunger began teaching her young blue heeler, Stella, to 'talk' using the same interactive soundboard she deploys to help young children master language (tapping a button on the board will play a recorded phrase). In one video, Stella taps the 'outside' and 'look' buttons in response to a sudden noise outside.

Copycat 'talking dogs' have since flooded the internet. A sheepadoodle named Bunny may have even achieved some kind of self-awareness – she was recorded playing the words 'Who' 'This' before staring at herself in the mirror. University of California's Rossano is now studying Bunny and 1650 other dogs learning to 'speak' with the soundboards at home – as well as 85 obliging cats and three curious horses.

'We're trying to work out if the [animals] are tapping the button because they've been trained to do it, because they get a reward or attention, or because they really understand,' Rossano says. 'Videos on Instagram can be edited. What we're looking for is if they start showing flexibility the way a toddler would, using a word like 'where' to ask for the location of different things.'

So far, Rossano says the canines in the study have outstripped his expectations – at times asking for help if in pain or even speaking for other dogs in the house the way an older sibling would. 'I thought the humans would be leading the conversations, instead we see the dogs initiating it. And, interestingly, most of their owners say they seem less frustrated now, they're barking less. It's kind of like a toddler screaming if they're not being understood.'

While Rossano is making no big claims until the data is in ('it may just be wishful thinking'), if animals do pass the test it could transform our relationship to them. 'Dogs seem particularly well suited to this [given] they're already so co-operative with humans. But you see pigs act almost like dogs if raised with people . . . what if pigs could [speak] too?'

Masson, Rossano and Vitale don't think it's unethical to keep pets, especially if adopting a stray. ('These animals are here now,' Vitale says.) But they stress that the more we understand the richness of animal thinking, the more we may need to rethink how we treat them.

What do our pets know of death, for instance? Dogs seem to want us with them when they are dying. 'Vets will tell you they panic if [their owners] leave the room when they're being put down, like they know,' Masson says. And cats often slink off somewhere, as if to wait out death alone.

Our old family cat slipped away into the garden like this the very evening I was finishing this article. She was always in the garden, our Padfoot. She liked to sneak up on unsuspecting visitors and start a game of tag. You'd hear a sudden, startling meow behind you, part greeting, part battlecry, and then a white paw would whip out from a bush to tap your foot, and the chase was on.

So we buried her there, under greystones from the pond. We lit candles and said a few words. She looked tiny in death; one of the finest hunters in the animal kingdom, wrapped in a seemingly unnecessary amount of fluff.

It struck me then. We may not always know what our animals really mean to us: are they roommates, workers, critics or family? They are not people, of course. But they seem to reach us somewhere other humans can't. And they leave us far too quickly.

> *We may not always know what our*
> *animals really mean to us: are they*
> *roommates, workers, critics or family?*

Whether they get as much out of this deal of companionship as we do, no one can say. But, millennia on, they are still at our side. And it didn't feel nearly as strange as I thought it would to give a cat a funeral.

28

HOW DO YOU FIX A BUSTED BALLET DANCER?

A snapped tendon could once mean curtains for a career. How do today's dancers make it back on stage when disaster strikes?

Nick Miller

You may imagine a serious ballet injury as a sudden, calamitous drama; a catastrophe in tights. It does happen.

When New York City Ballet star Robert Weiss snapped his Achilles tendon during a 1978 performance, the loud 'pop' famously carried to the stalls. He thought the floorboards had given way beneath him but, in the audience, his friend Rudolf Nureyev knew better and rushed backstage to help him to hospital.

Then there was American David Hallberg, the international superstar, alumnus of the Bolshoi and American Ballet Theatre, and now artistic director of the Australian Ballet. At his 2018 debut with the Royal Ballet in London – an eagerly anticipated moment long delayed by a career-threatening ankle injury – disaster struck.

'Midway through the first act, taking off for a jump, I felt a pop in my calf and from then on a sharp pain inflicted by just stepping on it,' he recounted on social media. 'I finished the act through sheer grit of determination but I knew in my gut I couldn't possibly continue.'

Career-threatening injuries can creep up on you, too. In 2016 one of the Australian Ballet's then rising stars Benedicte Bemet had seized the rare chance to dance a principal role, at the end of a busy season. 'I was in a bit of pain but I thought it was manageable,' she says of a niggly Achilles tendon. Fast-forward a couple of months and she could barely walk. 'I took my Christmas holiday in a bit of pain. And tendons notoriously hate rest. I didn't realise that at the time, I thought rest was good. It was really, really angry.'

Back in the 1970s, Weiss had surgery (luckily for him, a Danish specialist had been in the audience and supervised his operation). But operations are risky and can leave a dancer permanently restricted. For Hallberg and Bemet, recovery was slow and complex – though non-surgical – requiring huge reserves of physical and mental strength.

Why do ballet dancers get injured so much? And how is it that dancers who once might have had to hang up their pointe shoes forever are now returning from injury stronger than before?

Is injury common among ballet dancers?
Yes. But perhaps not for the reasons you might think.

A study in Britain in 2014 found that professional dancers were far more likely to suffer injuries than rugby players: 80 per cent of dancers incur at least one injury a year that affects their ability to perform, compared to 20 per cent for rugby or football players. Muscles and joints were the most common sites for injury, according to the British Fit to Dance 2014 survey. Other studies found that over-use was the most common cause of injuries for female dancers while men were more susceptible to sudden, traumatic injuries. And they found that younger dancers were more likely to be injured than older ones.

80 per cent of dancers incur at least one injury a year that affects their ability to perform.

Why are professional dancers so vulnerable? Partly because their workload is so intense: top companies perform many times more each year than sports teams play matches. There are 23 rounds in a typical AFL season, for example; in 2019 (a typical year) the Australian Ballet scheduled 289 live performances.

But Matthew Wyon, professor of dance science at the University of Wolverhampton and one of dance science's leading experts, has another theory. And it's not a flattering one.

'There's quite a bit of evidence that most ballet dancers are only slightly fitter than the average person on the street,' he says. 'I always call them "boy racer" cars: flashy exteriors but with an engine that is working overtime.'

There's quite a bit of evidence that most ballet dancers are only slightly fitter than the average person on the street.

One of the ways to measure fitness is the amount of oxygen you consume per kilogram of body weight: it's how efficient you are at using the fuel in the air to burn in your muscles. Distance runners and cross-country skiers would score in the 70s and 80s, soccer players in the 50s and 60s. Sedentary people would score low 30s. Dancers score from the 30s up to early 40s.

And, just as surprisingly, they're not that strong, either.

'We had a principal dancer just coming back from an ACL [anterior cruciate ligament] injury,' says Wyon, 'and we were testing his good leg against his injured leg, and the force [the good leg] was putting out was about 35 kilos. That's nothing.'

For comparison, an average gym bunny would barely raise a sweat doing a leg press of four or five times this amount.

Why is this so? Wyon believes it's because of the way dancers train. 'None of their training causes them to get either stronger or fitter until right up close to a performance. Ballet dancers are technically unbelievable. They've got an economy of movement we never see in sport. But it means the dance no longer puts a stress on the body. They don't have that physical adaptation. So, in fact, the better your dancer is, the less fit they are. Because dance doesn't stress them any more.'

On the face of it, the lifts and jumps that dancers perform seem to require extraordinary strength. But, behind the scenes, a lot is accomplished by perfect balance; by aligning bones and locking joints so that, rather than relying on muscles to hold your partner aloft, the weight transfers through your frame to the floor.

And the 'floating' appearance of a jump relies, again, on technique as much as lift-off power: by lifting and lowering arms and legs during a leap, the dancer's centre of gravity can travel the parabola that the laws of physics require while their body seems to defy science in mid-air. The same effect can be seen in a basketball player's hang time on their way to a dunk.

Evidence of their reliance on technique can also be found in dancers' almost freakish ability to ignore fatigue when it matters.

In one experiment, Wyon's team made a dancer exercise until they were 'absolutely dead on their feet' and then perform a double pirouette on to arabesque (which is where they stand en pointe with one leg in the air behind). 'And they could pull it off, even when they were having trouble doing the fatiguing dance in between. As soon as they were being watched, or having the data collected, they could pull it out. This is just a phenomenon and we're trying to explain it – and it could be how they're trained.'

Technique, it seems, honed over hours of practice each day and since an early age, hides a multitude of flaws. Wyon has seen a male dancer 'built like a stick insect' who could lift any of the women in the company – purely through ability. 'His technique was so good for doing it, beautifully. Once. But if you asked him to do it three times, he couldn't.'

> *Technique, it seems, honed over hours of practice each day and since an early age, hides a multitude of flaws.*

But why does this matter?

'They're always training and dancing at close to their maximum,' says Wyon.

Most of us operate around 40 per cent of our physical capacity in daily life. Sportspeople and dancers work at a baseline of around 80 per cent when they're 'on', occasionally pushing themselves to the limit.

'So, they're always at the edge rather than having any reserve,' Wyon says. And once fatigue sets in, 'that's when the potential for injury suddenly comes. Because he hasn't got that reserve underneath, to be able to protect him when he's tired.'

One wobble, one waver, and the dancer suddenly has to rely on a tired muscle or tendon that can't take it.

'They're really good at what they do but as soon as they get slightly outside that [they lack] physical competency, the ability to do other sorts of movements.'

This explains, says Wyon, why you tend to see an increase in injuries in dancers about a week, or two weeks, into performance time. Their bodies just haven't been prepared for the demands.

Injury prevention was not considered, supported or prioritised in the tradition-bound companies. Dancers would often not report niggles or early signs of injury, and would address symptoms only when they stopped them from performing. And the stakes are high. Until recently, injuries could easily end careers. Across the industry there was a view that dancers would perform until, at some point, they broke down . . . and that was that.

Are ballet dancers dropping like flies, then?

When the Australian Ballet's Damien Welch retired from the company he reported a laundry list of accumulated injuries: six operations on his legs and two screws in his foot (which he'd fractured during jumps), back injuries and stress fractures.

He was the son of Australian dance luminary Marilyn Jones, who herself damaged her Achilles tendon when a floorboard broke under her jump.

But the Australian Ballet is one of a group of pioneering dance companies around the world that have beefed up their in-house medical expertise and are leading the way in the search for better treatment, rehabilitation and – most importantly – injury prevention.

Dr Sue Mayes is the director of artistic health at the ballet, where she's worked since 1997 – at first in the littlest room in the building as the company's first full-time touring physio, now leading a high-tech medical and physiotherapy operation. A dancer in her youth, as she grew up Mayes was increasingly drawn to medicine, poring over the intricate, beautiful drawings of muscles, flesh, organs and bones in *Gray's Anatomy*.

Mayes sees all sorts of injuries, some unique to ballet. Pointe shoes lead to impingement at the back of the ankle, and pain and irritation to the structures around the bone. She used to see a lot of calf tears, as dancers pushed off from a jump and felt that tell-tale twitch in the muscle.

Knee injuries aren't as common in classical ballet as they are in contemporary dance, which has more twisting and turning moves while a dancer is planted on the ground. Around 30 per cent of injuries in classical ballet are to the lumbar spine: in women, due to the impingement of soft tissues in the lower back from bending into moves such as arabesques; in men, from the heavy lifting, catching, flinging and twisting of their partner. Hip problems make up around 7 per cent of their injury list.

Half the injuries Mayes' team sees are to the foot and ankle: swollen joints where friction and loading have led to inflammation.

'It's really common for dancers and athletes,' Mayes explains. 'You get the swelling in the joint and then the swelling can sort of escape, heading to the tissues, and they form these little balloon-like structures.'

Whenever Mayes puts a dancer under an MRI she finds a 'page-long list' of abnormalities around the joints, she says, though many do not affect performance.

But back injuries are much more dangerous. 'Over my time there were two or three reasons dancers had to finish [working], and one was injury to the lumbar disc,' she says. The other main reason was dancers having babies.

What can dancers do about injury?

There have been lots of advances in treatment of simple or niggling injuries but the more dramatic change is in the reduction of career-ending injury, and the reduction in the need for risky, all-in surgery.

Mayes recalls that when Bemet came in for treatment 'she had this tendinopathy around her heel. They can take a year [to mend]. They're really horrible. Every time you bend your ankle or put weight through your ankle it's horrible pain.'

Bemet had overloaded it. She had got to the state where she couldn't even plié – the knee bend that's the most basic ballet move. The fix was not cortisone or surgery. If you operate on something like this, the dancer will lose some range of movement in their foot, permanently. And that, in turn, can impact on other joints.

'We're [always] going to see if we can do it non-surgically,' says Mayes, 'because a dancer loves that swan neck, that hyper-extended shape. If you lose even five degrees of that it's going to be obvious to the eye and harder to function with. So, we avoid surgery at any cost – we've done very few operations in the last ten years.'

For a year, Bemet had to run through a simple, repetitive exercise routine involving the movement method Pilates, little jumps, or jogging up and down a stairwell, designed to restore strength and function to her foot.

It may sound simple, but in ballet it is a revolution. Rather than rushing dancers to hospital, they are experimenting with techniques to painstakingly rebuild the dancer from the inside out. Research published by Mayes and her team looks at each joint and each injury, and assesses what particular types, frequency and power of exercise are best to get a dancer back to the stage.

But without the perceived 'quick fix' of surgery, dancers must call on deep reserves of mental strength and patience. For four months, Bemet hardly improved. All she had to show for the strength exercises was increased strength; no reduction in pain.

'We just make these little short-term goals,' says Mayes. 'Rehab is this torturous roller-coaster of emotions and physical capabilities.

'A dancer is nourished by performing on stage; if they lose that opportunity they lose the opportunity to express their artistry. Their careers are relatively short and there are always younger dancers progressing through the ranks. If a dancer misses a year they may miss a performing or promotion opportunity.

'In the past, their identity was solely a "ballerina" due to growing up and spending their whole lives in a ballet school or company. In a long rehab, they can lose a sense of themselves, they lose their identity, they risk having nothing if they do not recover from the injury.'

And a long rehab is draining. It's hard work maintaining motivation, and a dancer can lose hope.

'They can start to question whether it is all worth it,' says Mayes. 'They question whether they will ever be able to dance again. If they return will they be as good? They miss the camaraderie. They can feel isolated.'

In a long rehab, they can lose a sense of themselves, they lose their identity.

Bemet says that dark time is 'still a part of me'. 'Until that point, ballet was how I identified myself, it was where I put my self-esteem, my value, my self-worth. To have that taken away . . . I was just completely lost, and quite depressed, especially because it really took such a long time to even get any sign that it was going to heal.'

In a memoir, Hallberg (who had the same injury as Bemet, exacerbated by botched surgery) reveals his own psychological crisis as he was working on rehab under Mayes' team. During his recuperative exile in Melbourne he spent hours on park benches, downing six-packs of Carlton Draught, 'the stereotypical drunk, the one whom everyone would fear and take pains to avoid'. He shaved his head and chain-smoked, sleeping in until noon.

But a common feature of great ballet dancers is determination. They didn't get to the top by chasing shortcuts.

'It was gruelling and time-consuming and meticulous,' recalls Bemet. 'You have to do [the exercises] at the same time every day, for the same amount of time, in the same order. It's so full-on. But I'm a bit of a perfectionist. I was more than happy to do everything they said to do. It was swallowing my pride and my ego, and knowing that I had to start back again from scratch, learning how to stand in first position, learning how to plié again, all these things I've totally taken for granted.

'You have to go back to baby ballet and be that blank canvas, take on information about your technique. But when you're so broken, you'll do anything. If they said, "Stand on your head", I'd be like, "Absolutely, I will do it."'

She came back a better dancer, and mentally stronger – the experience renewed her love and commitment to the art – but she was physically stronger, too. She was no longer a boy racer.

But don't ballet dancers hate looking muscly?

Mayes is still fighting battles to get her ballet dancers fitter and stronger, to ward off injury. One of the battles is against dancers' fear of strength. 'They don't want to bulk up,' she says. 'We want this ballet aesthetic that is long and slender.'

It's taken a long time to prove that a stronger dancer doesn't end up looking like the Hulk. It's all in the way you do it: Popeye muscles develop from high load and lots of repetitions, while the dancers work either with high load and low reps, or low load and high reps, resulting in muscles that can support and protect joints without bulging out of their leotards. It also helps that Mayes can prove to them they will get injured less.

Wyon says new research shows that a good, strength-based prevention program, based on techniques developed by Mayes and others, almost immediately reduces chronic injuries by half. Plus, they've done blind studies, with observers who don't know which dancer has been doing a strength program, and proven that fitter dancers dance better.

Another fight Mayes is still trying to win is against stretching – especially the unfathomably bendy stretching that dancers are able to pull off, whether they should or not. 'It's still kind of controversial,' she says. Ballet dancers love to sit in splits for hours. They wear it as a badge of honour, plus it feels good, particularly to a body battered by the demands of their career.

But, says Mayes, you don't actually want a too-stretchy calf, for example. When a joint goes to its full range of movement without the muscles to control it, that's when injury happens. Plus, research shows that stretching a muscle actually reduces its power output.

Mayes points out this is a lesson that Olympic sprinters have learned, too. 'You used to see sprinters stretching their hamstrings before the race and, almost inevitably, one of them would tear a hamstring during it.' But in recent Games they have switched to dynamic warm-ups: jogging, keeping their body warm. 'Immediately pre-race it's about keeping their heart rate up, all their muscles warm and their body active, but without any extreme range of movement.'

Younger dancers especially have to be trained out of stretching. Wyon finds himself spending hours on Instagram telling young dancers (or, often, American cheerleaders) who love to do extreme stretches, 'that's bad, you shouldn't be doing that'.

'They're trying to force their bodies into these weird and wonderful positions but they're stretching ligaments rather than muscles and tendons,' he says. 'That means we've got increased laxity, which means we are more likely to get injured, unless we increase the muscle strength around that joint to be able to control it and support it. You get these cases of wibbly-wobbly knees, and when they get fatigued they won't be able to control their bodies.'

Do dancers stay fixed long-term?

For more than 15 years, since it transformed its injury treatment and prevention program, the Australian Ballet has not lost a single dancer to injury. 'Which is amazing,' says Mayes. 'We've got women who are dancing into their late 30s, men dancing until they're 40, which was never heard of in our company.' When they leave the company now, it's rare for a dancer to leave with an injury that bothers them.

Mayes also hopes the work they're doing will help dancers long after their retirement. 'Dancers that retired before we really brought in all the strategies are getting hip replacements in their early 50s. We won't know [if we've changed that] for another ten years, but I'm hoping.'

She can now look her dancers in the eye and tell them that, whatever it is, she can fix it. 'The thing we always start off with is, you'll come back from this injury. And this will not just leave you where you were at. If you commit to this rehab, you will come back better than ever.'

And Bemet did. She fought back, she healed, she returned to the stage. And she was promoted to principal artist soon after.

29

WHY DO RUSSIANS GET POISONED?

The Kremlin's foes have a much higher
chance of succumbing to rare poisons
than the general population. Why?

Sherryn Groch

There's a saying in Leonid Petrov's native Russia: Drink the vodka, not the tea. 'Russia's politics can be toxic,' explains Petrov, an expert on Russian and Korean history now based in Australia. And the Kremlin's revenge is often served piping hot.

Former spy Alexander Litvinenko was famously murdered with a cup of radioactive tea. Journalist Anna Politkovskaya drank a laced brew while flying to cover a crucial story, and instead woke up in hospital. And in 2020, two years after the Russian military toxin Novichok was unleashed on double agent Sergei Skripal and his daughter, Yulia, on British soil, it was deployed again against another high-profile enemy of the Kremlin.

This time the target was popular opposition figure Alexei Navalny, a man many say was beginning to outsmart the sprawling apparatus of the Russian state just as a new mood of rebellion swept through neighbouring Belarus. On 20 August, Navalny boarded a flight from Siberia to Moscow. An hour later, he had to be carried, screaming, from the aircraft bathroom as the aircraft made an emergency landing.

Political poisonings may seem like the stuff of medieval intrigue or Cold War legend (take the 1978 case of the Bulgarian journalist murdered with a poison-tipped umbrella on a London bridge, for example). But poison has remained a signature weapon of the Russian state since Moscow's secret poisons laboratory No. 12 began experimenting with chemical agents in 1921.

Today, critics of the Kremlin still have a much higher chance of succumbing to rare poisons than the general population – although the Russian government has consistently denied any involvement in the string of suspicious deaths and illnesses that have befallen politicians, spies and journalists down the years.

In the case of Navalny, calls from world leaders for a proper investigation (or explanation) have again been brushed off by the Kremlin. So Navalny, recovering from the attack in Germany, took matters into his own hands – releasing a recording in which he

Why do Russians get poisoned?

apparently duped one of the FSB (Russian secret service) agents on his tail into admitting how they poisoned him (using Novichok sprinkled on his underwear).

A laughing President Vladimir Putin later admitted his agents had been tailing Navalny but said if they had wanted to kill the critic, they 'probably would have finished it'. Navalny was not cowed. In early 2021 he made good on his vow to return to Russia ahead of its elections. He was swiftly imprisoned, declaring from the dock that Putin would go down in history as 'Vladimir the underpants poisoner'.

So why do Russian dissidents keep being poisoned? Can anything be done about it? And what does the Navalny case reveal about the state of Russia – and Putin's hold on it – today?

Why poison?

Dr John Besemeres has forged a long career as an intelligence analyst and Russia expert and even he 'can't keep track of the dozens and dozens of murders' and attacks on political dissidents in the past two decades. 'It reflects the KGB tradition: "If anyone betrays us, we kill them, we follow them to the ends of the Earth",' Besemeres says. 'Putin is ex-KGB, and he hates traitors.'

The appeal of poison for Russian assassins is twofold, according to a man who has lived much of his life on the wrong side of Putin, US financier and human rights campaigner Bill Browder. Poisons are notoriously difficult to trace and deaths are sometimes blamed on a victim's existing health conditions. 'On the other hand, everyone knows who did it,' Browder says. Poison has become a kind of Kremlin calling card. 'Putin likes to have it both ways ... make a symbolic point but [escape] the consequences,' he says. 'The message is clear: if you challenge [him], you will die.'

Poisoning is also typically painful, and recovery – or death – is slow, making it viciously theatrical. Two of the most shocking poison attacks in recent memory – against former spies Litvinenko in 2006 and Skripal in 2018 – left a toxic trail through Britain and the victims in hospital for weeks. While the Skripals survived, one woman died after her partner gave her a perfume bottle he'd found in a charity bin, which, unbeknown to him, contained the Novichok. And Litvinenko suffered a slow, agonising death in a hospital bed, 'a clear warning to others like him', Besemeres says.

Writing of her own poisoning in 2004, Politkovskaya said the choice for Russian journalists had become 'total servility to Putin' or death. 'It can be . . . the bullet, poison, or trial – whatever our special services, Putin's guard dogs, see fit.' Two years later, on Putin's birthday, Politkovskaya was shot dead in an assassination blamed on Chechens but which, again, cast suspicion on the Kremlin.

Browder says the mysterious misfortunes of Kremlin critics have increased in the 20 years that Putin has been in power. Among the arsenal of toxins identified in recent incidents are the radioactive isotope polonium-210, a rare Himalayan plant toxin and, of course, Novichok, the Soviet nerve agent lethal to the touch. 'Really exotic stuff,' Canberra emergency doctor David Caldicott says. 'These are some of the most lethal chemicals humans have ever created.'

In Navalny's case, strange details quickly emerged, such as the mysterious (and fake) bomb threat that forced an evacuation of the airport where Navalny's plane was making an emergency landing. Or the Russian doctors, who insisted officially Navalny was suffering from a 'metabolic disorder' and so blocked his transfer to a Berlin hospital for more than a day – 'almost the exact amount of time the poison would need to leave his system', Caldicott notes.

When Pyotr Verzilov of the Russian protest group Pussy Riot fell violently ill in 2018, in circumstances he now says eerily mirror Navalny's, he was kept sealed off in a Russian ICU ward for days before being released to Berlin. In his case, it was too late to pinpoint the exact poison. In Navalny's, a German military lab found 'unequivocal proof' of Novichok.

Such a poison can interfere with the crucial neural pathways in the body that control breathing, heart rate and digestion. Treatment is dangerous and difficult, both for the patient and staff due to the risk of contamination, Caldicott says. Litvinenko's autopsy is still considered the most dangerous post-mortem ever conducted in the modern world. Navalny's own entourage were at first told he may be too dangerous to approach without protective gear. He remained in a coma for almost three weeks.

Who is Alexei Navalny and why was he a target?

In a country where elections are heavily manipulated and genuine opposition candidates do not see the inside of parliament, lawyer and anti-corruption campaigner Alexei Navalny has used the internet to mobilise considerable swathes of the population and disrupt Putin's propaganda. Since politician Boris Nemtsov was shot dead just metres from the Kremlin in 2015, the charismatic Navalny has been the de facto head of the opposition – and Putin will not even speak his name.

His slick video exposés and commentary draw millions of views and target Putin's inner circle – including the powerful oligarchs who took control of Russia's resources after the fall of the Soviet Union. For a 2017 investigation into the vast empire amassed by Russia's then prime minister Dmitry Medvedev, Navalny's drones flew over palaces, yachts and secret dachas (or seasonal houses) across the country – all built with 'donations' from oligarchs and banks.

The scale of the wealth he uncovered gave Russians a rare glimpse into how they were being ripped off, Petrov says. One of Medvedev's country estates, Navalny quipped, even appeared to have a house built just for the ducks on its lake, and ducks quickly became a symbol of mockery and protest, turning up in inflatable form at rallies even beyond Russia's borders.

Nemtsov before him had also exposed Putin-era corruption. The former deputy prime minister had been preparing a report detailing how Russian troops were secretly fighting in Ukraine when he was assassinated. More critically, Petrov says, Nemtsov was preparing for another run at parliament, and from there, a presidential tilt. 'Putin couldn't permit this,' Petrov says. 'After he was killed, Navalny stepped in [as the main] opponent.'

While Navalny was himself disqualified from running for president under a trumped-up embezzlement conviction in 2014, Besemeres says 'He Who Must Not Be Named' has still become a serious problem for Putin. He has turned the formidable Russian government into the 'Party of Crooks and Thieves', a foe that can be weakened, perhaps one day even overcome, through his tactic of 'smart voting' that encourages Russians to vote for local representatives in regional elections over the official Kremlin candidate at all costs.

This plan saw Putin's Russia United party lose a third of its seats in the 2019 Moscow election; and a local, Sergei Furgal, elected to the role of governor in far-east Khabarovsk. Furgal's subsequent arrest by the Kremlin on murder charges was widely considered to be politically motivated and sparked months of protests – then the biggest unrest Russia had seen since Putin took office.

But the Furgal protests were quickly surpassed by the tens of thousands who took to the streets in the bitter cold all across the country to chant Navalny's name as the critic was thrown in a Russian prison in early 2021.

'In 100 years' time, the history books in Russia will discover that the most important person of the last two decades was not Putin who disgraced us but Alexei who saw above it all but took terrible risks,' Besemeres says. 'They must be consumed with hatred for him in the President's office.'

La Trobe University's Russia expert Dr Robert Horvath agrees that if a democratic Russia one day emerges 'from the wreckage of Putinism', Navalny will have been its prophet.

Navalny had had warnings even before his poisoning – he was twice doused with a corrosive antiseptic on the street in attacks that left his skin dyed green and damaged his eye. He also fell ill with a mysterious 'allergic reaction' when he was previously locked up by the state for organising unsanctioned protests. His brother was put under house arrest and his anti-corruption fund was blacklisted 'as a foreign agent', although Navalny says he does not receive overseas money.

'When I was watching Navalny's latest broadcast [before the poisoning], I was wondering, how can he still be alive?' Petrov admits. 'He was calling Putin a thief, a criminal – not just the party – and he'd been linking the corrupt officials in his investigations to him more and more. The President is very sensitive to these kinds of personal attacks.'

Even after the 2020 poisoning, as Navalny sat in prison, he released his most explosive exposé yet: claiming that Russia's oligarchs had secretly built Putin an extravagant billion-dollar Black Sea palace, which Navalny called 'the largest bribe in history'. The Kremlin denied the report. Putin's former judo sparring partner claimed the estate – which boasts its own casino, vineyard and underground ice-skating rink – belonged to him instead.

Still, Navalny's survival in Russia, despite regular police raids and fines, had made many see him as almost untouchable – someone to be worn down by the Kremlin with frequent but relatively short prison sentences, not martyred with a swift exit from Russian politics. In February 2021 he was sentenced to his longest stint yet: almost three years in a penal colony (one of the country's notoriously brutal prison camps) for breaking parole on those embezzlement charges when he was in hospital recovering from the poisoning.

▌ What does the Navalny poisoning tell us about Russia?

There may be no official rubber stamp from the Russian secret service on the attack, but the world is in little doubt about who is to blame. The United States, Britain and the European Union quickly imposed sanctions against Russia in response. Navalny himself has even heard the voice of his would-be assassin when he posed as a top security official in order to trick one of the FSB agents regularly on his tail into 'debriefing' him on the poisoning.

While Navalny has enemies beyond just the President, not least of all the string of corrupt officials he has exposed in his investigations, most experts scoff at suggestions the hit could have been carried out without the President's sign-off. '[Navalny] is the most well-known person outside Putin,' Petrov says. 'Without Putin's consent, it's hard to imagine someone would touch him.'

Besemeres agrees: 'It's Putin's regime, he's the boss, so there might be other people acting but the options really become, was it this arm of Putin's regime or this arm over here?'

And the timing may have been more telling than Putin would like – ahead of crucial 2021 elections and as the people of neighbouring Belarus, Russia's closest ally, rose up against their own long-time autocratic leader, Alexander Lukashenko.

Browder and Horvath say Putin fears revolution contagion back home. '[He was] panicking,' Browder says. 'The natural person to lead [an uprising] in Russia would be Navalny.'

Besemeres agrees the poisoning appears to be a 'desperate but emphatic' move on the part of the Kremlin. 'It's meant to make people's blood run cold,' he says. 'Let the serfs know they will not be talking back to the masters any more. Not many people are as brave as Navalny.'

Some fear it might signal something more – that Putin is done playing at democracy and will no longer entertain any real opposition in Russia. Certainly, record protests against Putin following Navalny's return and arrest in Russia were met with a notably brutal crackdown by authorities. Thousands were detained.

'It's a very scary development from that point of view,' Besemeres says. 'His behaviour lately has been getting more and more extreme, there've been more arrests, more people run out of Russia.'

Journalist Alexander Baunov has argued that if Putin is crossing the line from hunting down ex-spies such as Skripal and Litvinenko to a figure as prominent as Navalny, then it's a sign 'the regime – certainly its most hardline elements – feels more endangered than ever'.

While Putin has recently rushed through a suite of amendments to the constitution to solidify his rule (which would let him run for two more six-year terms, taking him to 2036), his popularity has slumped in the face of an economic crisis at home (even pre-COVID) and his controversial decision to raise the pension age.

'Navalny is someone who sensed the weakness of Russia as a federal state, who understood it's a colossus on clay feet,' Petrov says. As in Belarus, he says, the people are tired of a failing economy and government spending on 'guns not bread'. 'All Putin seems to do is fearmonger, claiming NATO is on the doorsteps and without him Russia will fall prey to a Western conspiracy. But as soon as people see the vulnerability of the state, they will start to question Moscow's legitimacy to rule, to collect taxes . . . It's what Putin fears most.'

Others say Navalny does not pose a serious challenge to Putin, especially now he is in prison. Such is the power of the regime that its end will likely come from within, not from a man calling protesters to the streets.

While Putin's strongman image, standing against Western foes, has become more tired and *uncool* of late, some commentators say it was given a shot in the arm by US President Joe Biden, who very pointedly branded Putin a 'killer' in early 2021. Such was the anger within Russia that an editorial in the newspaper *Kommersant* predicted Putin's party might win every seat in Russia's parliament 'thanks to Biden'. Meanwhile, very quickly, Putin, who has never debated a domestic political opponent in his life, challenged Biden to a live debate (not the done thing among world leaders). And his famed outdoor (and occasionally bare-chested) photoshoots have returned with gusto, as his propaganda machine goes into overdrive to shift the narrative away from unrest.

Ultimately, Browder thinks the attacks on Navalny will tip the scales one of two ways: 'either destroy the opposition in Russia or be the straw that breaks the camel's back and leads to uncontrolled uprisings. Much depends on what happens to Navalny.'

What is to be done?

While lower-tier hitmen involved in such attacks are sometimes charged, justice never quite reaches those who give the orders. In the case of the Skripal poisonings in Salisbury, the two Russian agents identified by British intelligence flew straight from their UK 'sightseeing trip' back home to Moscow, where they remain safe from extradition. Litvinenko's assassins, former KGB colleagues, also escaped British justice, even after a UK inquiry finally found, a decade on, that the hit had 'probably' been ordered by Putin.

'With Salisbury, something so shameless, even then the West has been weak and soggy in its response,' Besemeres says. 'A lot of these high-ranking Russians enjoy having a summer house in France or London and those countries enjoy their money.'

Browder's lawyer, Sergei Magnitsky, died in 2009, sick and reportedly beaten in a Kremlin prison, after discovering a $230-million tax fraud carried out by Russian officials. Four other witnesses to the scandal also died in mysterious circumstances.

In Magnitsky's name, Browder lobbied for the United States to create a new line of sanctions that target individuals involved in human rights violations and gross corruption, rather than whole nations, blocking visas and bank accounts and even seizing properties. Versions of this Magnitsky Act have since been adopted by a number of countries – most recently by all members of the European Union. Navalny is a big supporter of the act and his poisoning helped galvanise the EU's adoption of the sanctions regime. 'It seems particularly fitting that the Magnitsky Act should be applied to the people who tried to kill Navalny,' Browder says.

Of course, not everyone in Russia is conspiring to bump one another off. Browder, who lived in Moscow for a decade before he was deported for speaking out on corruption, says: 'Russian people are some of the most honest, brave and idealistic people you will ever meet. Take Magnitsky. He was thrown in jail and he still refused to recant [his testimony] . . . It's a shame it doesn't come across in their international reputation, in their government.'

Still, the shadow of FSB surveillance often demands precautions, even beyond Russia's borders. The champion chess player turned opposition figure Garry Kasparov reportedly has bodyguards oversee his meals and carry bottled water. And a professor who predicted the end of Putin's reign by 2022 recently confessed to independent Russian media that he had been warned his poisoning was already planned and he should steer clear of drinking tea or coffee.

Browder himself has been the target of death threats and kidnap plots as well as persistent 'red notices' from the Kremlin requesting that Interpol arrest and return him to Russia. If assassins do come calling, he imagines poison will be a likely weapon of choice.

'Russia is quite a dangerous place to be,' Petrov says. 'You never know who you are talking to and who's watching you. But overseas, things usually only happen if you are a particular value or danger to the Russian leadership, then special agents are sent to find a way. And they do.'

DRINK THE VODKA, NOT THE TEA

Poisonings of note

Vladimir Kara-Murza

The opposition activist was poisoned twice, once in 2015, shortly after the murder of then opposition leader Boris Nemtsov, and then again in 2017. Kara-Murza survived both incidents, which he believes were the work of the Kremlin, but says his doctors worry he won't survive a third.

Viktor Yushchenko

The former president of Ukraine was poisoned in 2004 with an 'Agent Orange' chemical served in a rice dish that left him permanently disfigured. He had been running for president against a Russian-backed candidate at the time but went on to win the vote and take office, later blaming the attack on the Kremlin.

Alexander Litvinenko

The former KGB agent had published a book from exile in Britain linking a series of apartment bombings in Moscow and other cities that killed more than 300 people to Putin and the Kremlin. In 2006, he fell ill after meeting with two former Russian agents and drinking tea laced with the radioactive isotope polonium-210. He died weeks later in hospital and a British inquiry eventually concluded Putin 'probably' ordered the assassination.

Anna Politkovskaya

The renowned Russian journalist had also reported on the apartment bombings and was on her way to cover a school siege in Beslan in Chechnya when she fell unconscious after drinking tea on a plane. She believed she had been poisoned by Putin's agents and two years later, on the President's birthday, she was shot dead in the lift of her apartment block, in an attack blamed on Chechens (who were also blamed for the apartment bombings).

Sergei and Yulia Skripal

The former spy and his daughter were targeted by Russian agents with the military-grade nerve agent Novichok in 2018. They were discovered on a park bench in Salisbury, England, where Sergei had been living. While they both eventually recovered, two Brits also fell sick from exposure to the chemical, one of whom died.

Pyotr Verzilov

A spokesman for the protest group and band Pussy Riot, Verzilov was rushed to Berlin for treatment in 2018 after a suspected poisoning. He recovered but blames the Russian government for the incident, which happened just weeks after Pussy Riot disrupted the World Cup soccer final in front of Putin and other world leaders.

Yuri Shchekochikhin

The Russian journalist died of mysterious symptoms consistent with poisoning in 2003, just days before he was to leave for the United States to meet with FBI investigators. His death was widely considered to be a Kremlin assassination.

30

HOW DO YOU FIGHT A CYBERWAR?

Hackers can stop the trams and turn off
the lights. But could they start a war?

Sherryn Groch

O n the morning of 27 June 2017 it seemed as if Ukraine had slipped back in time into a previous century. There were no ATMs, trains, airports, television stations. Well, they were there – but they didn't work. Even the radiation monitors at the old Chernobyl nuclear plant were down.

Ukraine, in the midst of a long and undeclared war with Russia, had been hit by mysterious blackouts before but this was eating through computer networks at a terrifying pace, turning screens dark across the country. And it was spreading further still, out through Europe and around the globe, paralysing hospitals and companies from London to Denver, even the Cadbury chocolate factory in Tasmania, and bringing swathes of the world's shipping to a halt. By the time the culprit – a wild variant of malicious computer code (or worm, or virus) known as NotPetya – was stopped hours later, it had looped back into Russia, where it originated, and racked up about $13 billion (US$10 billion) in damage worldwide, making it the most expensive cyber attack costed to date.

No one died but the world had glimpsed a new reality, beyond cyber espionage or sabotage. This was cyberwar. With modern life more connected than ever, you could unplug a nation before you'd even fired a shot.

Today, cyber weapons feature in the opening moments of most countries' war plans, but they are deployed in peacetime, too, and the line between espionage, vandalism and outright attack is far from clear.

In 2016, the Australian government broke its relative silence on the cyber threat, revealing for the first time that it was actively engaged in cyberwarfare (against the terrorist group Islamic State in Syria and Iraq) and warning of a coming cyber storm. The army's newest head of Information Warfare, Major General Susan Coyle, says it is now seeing an exponential growth in the range and sophistication of cyber weapons. Top companies and universities have been mined for personal data; media company Nine Entertainment, which owns *The Age* and *The Sydney Morning Herald*, was briefly knocked off air in Sydney in March 2021; even parliament itself has been infiltrated. But Coyle says Australia's cyber forces are being rapidly trained to meet the threat.

So, what would happen if the skirmishes of cyberspace did break out into real-world death and destruction? How vulnerable is Australia to the kind of attack that knocked the lights out in Ukraine? And is there a way to keep the great cyber powers in check?

What is cyberwarfare anyway?

In 1993, just four years after the world wide web first sparked to life, a US think tank warned 'Cyberwar is coming!' It was right.

In 2009, the world's first digital weapon was unleashed on a foreign state – a worm built by the United States and Israel that became known as Stuxnet. Its target was Iran. At 15,000 lines of code, Stuxnet was designed to do more than steal data or crash computers. Like any good spy, it learnt and lay in wait, feeding false information into the safety sensors at an Iranian uranium enrichment plant until one day it sent the site's centrifuges into an unstoppable, destructive spin. The plant was so damaged it set back Iran's nuclear program by months, likely years.

Only Stuxnet didn't disappear as planned; it got out, infecting thousands of machines across the world. While the worm is now dormant, programmed to come to life only in specific conditions (such as arriving on software at an Iranian nuclear facility), this military-grade weapon has been out in the open, in the hands of security experts, rival states and criminals since that initial attack. And experts say the game has only become more dangerous since.

The internet may be the great connector but the access it opens up into each of our lives has long been exploited by hackers – be they spies, saboteurs, thieves, activists or bullies. While this regular back and forth lends itself well to the war analogy, most of what goes on, even between nation states, still falls below the threshold of actual warfare. It lives in 'the grey zone', says Tom Uren, a former Defence cyber analyst now at the Australian Strategic Policy Institute (ASPI).

Stuxnet itself possibly prevented real war, blunting Israel's perceived need for a military strike against Iran developing a nuclear bomb. (And, a decade later, the Trump administration called off a planned strike on Iran in favour of a cyber attack.) 'The things done in the grey zone aren't always adversarial,' Coyle says. 'That's how you learn what threats are out there, what everyone else is doing.'

Still, the stakes are getting higher. The rise of artificial intelligence (AI), satellite technology and the internet of things (where more devices, from lights to door locks, are connected) means targets are opening up faster than we can patch vulnerabilities. China's state hacking teams steal corporate secrets, as well as government data to blunt the West's military advantage. Russia hijacks social media not just to spread propaganda but to manipulate democracies. And nations can turn these cyber weapons on their own citizens, too, to stamp out dissent.

'It's not warfare but it's definitely not peace either,' Uren says. 'Some countries will push right up to the edge of that red line using covert, deniable methods . . . cyber is perfect for that. NotPetya is probably the closest we've come to real war.'

NotPetya hit during an actual physical invasion, too – Russian troops (and bikie gangs) had already been sent into Ukraine without military insignia to seize Crimea and sow violence. Likewise in the former Soviet republic of Georgia in 2008, cyber attacks seemed to hit towns just ahead of Russian soldiers arriving to back pro-Russian separatists. The year before, when Estonia, one of the most wired nations in the world, was unplugged, it went to NATO for help. There was even (brief) talk of invoking

*It's not warfare
but it's definitely
not peace either.*

Article 5, which demands all other nations in the alliance defend one another from enemy assaults. But the world did not see a direct military retaliation to a cyber attack until Israel bombed a building linked to Hamas hackers in Gaza in 2019.

Ukraine has become Russia's testing ground for cyber weapons, as Taiwan is now for China, says Professor Greg Austin, a former government adviser and analyst who heads a program of cyber war study at the University of New South Wales. But, for all the Kremlin has unleashed, it's still holding back. 'It's not looking to crush the Ukrainian government entirely,' Austin says. 'In a major war, everything Russia is doing in Ukraine, it would do 100 times over to many more targets [elsewhere] ... And other countries have huge capabilities, too.'

New York Times journalist David Sanger has watched cyber conflict heat up since he first helped unravel the mystery of Stuxnet in 2012. As luck would have it, he even found himself in Kiev years later just as NotPetya was hitting ('I didn't have any Ukrainian money, and all the ATMs were down'). But he agrees the world has not seen full-scale cyberwar yet. Digital weapons are still mostly deployed as 'short of war' tools, he says, cheap, effective and often difficult to trace back to the state actor, making retaliation complicated.

Indeed, unlike regular weapons, cyber has become a tempting way for smaller nations to show their teeth without invoking devastating counterstrikes. Just nine countries have nuclear weapons but most have state-sponsored hackers. That means attacks can come from almost anywhere and, as many experts warn, could steer dangerously out of control. 'We are where we were with aeroplanes at the end of the First World War,' Sanger says. 'It's still mostly used for [surveillance] but the weapon is there.' And, once that line is crossed and countries are at war, then cyberspace, just like air, land, sea and space, becomes another domain in which to take out the enemy.

When do shots fired online count as acts of war?

After the cyber attacks on Estonia, dubbed Web War I, the question of what constitutes an armed attack in the digital age became live. Through NATO, academics drafted the Tallinn Manual, named for Estonia's capital, to lay out how international laws of war might apply to cyberspace.

Professor Dale Stephens, a lawyer and former Navy captain, helped peer review the Tallinn Manual and is now working on a similar 'user's guide' to international law applicable to military operations in space, known as the Woomera Manual. Space, he says, already has some well-established norms of behaviour. 'It's governed by five treaties. Russia and the US have already worked out their tolerances with each other.'

> *Cyber has become a tempting way for*
> *smaller nations to show their teeth without*
> *invoking devastating counterstrikes.*

In cyberspace, he says, countries are still very much feeling out those boundaries. Under the laws set out in the Geneva Conventions and other treaties, blowing up a rival nation's battleship is clearly warfare. 'But suppose I take my ones and my zeroes [of computer code] and I manipulate your battleship's systems until it's damaged, or it blows up,' Stephens says. 'At what point then am I crossing the line?'

Is the malicious code that both Russia and the US now implant in each other's power grids, for example, just routine surveillance or the first act of a devastating strike? What if the pacemaker of a foreign leader was hacked? Even social media itself can be weaponised to gain a military advantage. During the 2014 Islamic State campaign in Iraq, a carefully orchestrated jihadist storm online (featuring horrific videos of executions and overblown claims of victories) convinced the 25,000-strong Iraqi garrison that they didn't stand a chance against the terror group. In reality, IS fighters in the area numbered only about 1500. 'The Iraqis surrendered and gave IS [the city of] Mosul,' recalls Stephens, who was serving elsewhere in the region at the time.

'Most people can agree on the big stuff that's crossing the line. But then there's the stuff just below, where they're using our systems against us in a kind of information war that [fractures] a state; that can be as threatening as destroying those systems entirely. Some of what's going on may already be a use of force [under international law]. But what's a proportional response? . . . Even the Tallinn Manual is still just recommendations.'

'There are some things below an armed attack [in the law] which are still nasty,' adds Austin, recalling the deadly 1985 bombing of a Greenpeace ship in New Zealand by French intelligence agencies. Years on from the Estonia hack, NATO now says it will invoke Article 5 in the event of a serious cyber assault against an ally (the mode of retaliation depending on the severity).

In 2019, Australia solidified its own position: when a cyber attack poses an imminent risk of damage equivalent to a traditional armed attack, such as significant loss of life or critical infrastructure, then a country should be able to defend itself. France and Denmark have spoken of their right to sovereignty, not just safety, in cyberspace. The United States has left the door open to taking some extraordinary steps, even nuclear ones, against a serious cyber attack – and has loosened the reins on US Cyber Command, allowing the military to launch some strikes without presidential approval in the same way they do in other theatres of war. It's part of a modern US strategy on cyber, which Australia, as a member of the Five Eyes intelligence alliance, is also following to some degree, known as 'defend forward'.

Austin explains: 'That means if China or Russia are persistently trying to penetrate our systems we're going to stop them even if it means going into theirs.'

Coyle, who was the first woman to command all of the Australian Defence Force's operations in the Middle East, can't offer much detail due to 'the classification level' of the cyber operations she now oversees. But she says everything the ADF does complies with the law, both in Australia and internationally, and commanders take pains to make sure no one steps outside those boundaries.

Who are the big cyber powers at play?

The online world looks a lot like the offline one – the United States, China and Russia remain at the centre of power struggles. America is still considered to have the most advanced cyber capabilities in the world. But China, Russia, Israel, Britain, even Iran and North Korea also have formidable cyber armies – think of the legions of hackers installed in St Petersburg or behind China's great firewall.

Still, some countries are noisier than they are effective, Uren says. 'Often a really great, well-executed operation you don't know about.'

Russia and North Korea are conspicuous in cyberspace for the same reasons they are on the world stage: shows of force. Here they use digital weapons not just for espionage and war but political point-scoring, even harassment. Remember North Korea's attack on Sony Pictures in 2014 ahead of the release of a comedy critical of its leader Kim Jong Un? Or the hacks that paralysed broadcasts of the 2018 Winter Olympics after Russia's doping scandal (these were even codenamed Sour Grapes by intelligence agencies linking them back to Russia)?

Austin has been analysing the cyber arsenals of countries and says that the smaller nations making headlines, such as Iran and North Korea, don't have the same depth of capability as the big players. 'They can still cause damage but they're not able to launch something as sustained and wide-ranging. The big ones could shut us down, close off traffic lights, stop the trains running, and make it last longer.' Still, these smaller nations consider they have one big advantage – they are not as wired as their Western adversaries, making their own exposure smaller.

In Australia, most attacks considered sophisticated enough to be attributed to another state are thought to have come from China, although the country has denied them, as it does all hacks. China has been less brazen than Russia in its cyber attacks on the West, mostly sticking to espionage so far. Still, its restraint does not extend to Taiwan, where cyber attacks come almost daily. And, as diplomatic and trade disputes escalate with Western countries, notably Australia, some fear China is growing bolder. In March 2020, an attack crashed the website of a global coalition of MPs speaking out on China's aggression.

> *China has been less brazen than Russia*
> *in its cyber attacks on the West,*
> *mostly sticking to espionage so far.*

'China has learnt from Russian interventions in elections in the US and Europe,' Austin says. 'They can do more things in cyber than they previously imagined.'

Stephens adds that the growing superpower is also investing heavily in new technology that will shape the future cyber battlefield such as AI, satellites and 5G networks. 'China's trying to jump that industrial step the West took of building big fleets of ships and airplanes and go straight to the next generation of weapons: cyber and AI.'

Back home, experts agree Australia is at last taking cyber more seriously, recruiting more hackers and rolling out new cyber security standards to shore up privately owned critical infrastructure. But, while we are not exactly trailing the pack on cyber security internationally, we are still not doing enough. 'Look at what our adversaries are doing,' Austin says. 'You see our big government departments starting to uplift their security and still only put in a moderate performance. And they're only transparent [about attacks] when it suits them.'

America's Cyber Command is thousands strong, created in 2009 after a particularly embarrassing breach of Pentagon internal networks by the Russians. Australia didn't have its own military cyber force until 2017, Coyle says, and now has about 400 personnel, a mix of soldiers, contractors and public servants who work within Defence and sometimes with Australia's spy agencies. 'We've come a long way very fast but we're still learning,' Coyle says.

Still, Austin says, the West (and specifically the United States) is winning the cyber battle. 'The broad narrative that China is winning is really a gross exaggeration; their cyber defences are weak,' he says. 'And we never hear of all the times the West successfully hits them or Russia.'

But how likely is a cyberwar and how bad could it get?
To get a full-scale cyberwar, where nations are actively unplugging their enemies, experts say the world would have to be either already on the brink, or an attack would have to spiral rapidly out of control, into something interpreted as a clear act of war. Uren imagines it would take a big attack 'something with the impact of [almost a] 9/11 where you had mass casualties, not just mass destruction of IT systems'.

We haven't seen that yet. And while geopolitical tensions have only escalated during the COVID-19 pandemic, the superpowers remain reluctant to go to war. 'Even calling an attack warfare means you have to respond,' Uren says.

Coyle adds: 'I think we'd need some pretty incredible evidence to suggest something was an act of war, that it wasn't an unnecessary escalation or a mistake. And I'd be surprised if somebody was stupid enough to want to do that, knowing that, collectively, countries would go against [them].'

But, while she is less concerned about one strike taking many lives – the doomsday 'cyber Pearl Harbor' scenario – she says even a hack causing mass disruption, such as knocking out power, could hit with a force akin to a natural disaster. People could still die. 'And economies can fail. We've seen it with COVID. Things can change quite rapidly. If we were to be attacked ... Australia-wide, the impact would be far reaching.'

> *A hack causing mass disruption,*
> *such as knocking out power, could hit*
> *with a force akin to a natural disaster.*

Stephens says the greatest threat may come from attacks above, with cyberspace increasingly connected to satellites. GPS doesn't just help you find where you're driving and video chat to people on the other side of the world; it's integral to military operations. 'We can restore our systems down here but if I take down the satellites that help them run, that's going to have a much bigger impact,' Stephens says. 'We've just woken up to this vulnerability.'

Coyle agrees an attack on space infrastructure would be very concerning but stresses, 'I'd hope that we don't have one point of failure ... We still use paper, for example, we can use compasses if space fails.'

As Stephens puts it, 'cyber ends at a certain point'. He doesn't imagine it will ever pack the kind of knock-out blow of a nuclear weapon. 'There's always a patch, there's always a defence. I think the US has huge capabilities to unleash a devastating cyber retaliation. But the world will survive it. Of course, AI might change that. If I'm on an aircraft carrier on the South China Sea and I'm suddenly swarmed by a bunch of self-driving underwater drones, I'm not standing a chance.'

Austin agrees the marriage of AI with weaponry could ratchet up the stakes in the coming years. And Uren says that, while a 'cyber Pearl Harbor' is unlikely, 'with cyber it's difficult to rule anything out'.

'It's hard for me to imagine we'd get a first-strike capability that could disable another country's military but ... if you could switch off the air defence radars [of another nation], for example, you could just fly in your bombing planes.'

Is cyber war inevitable?

In his 2018 book *The Perfect Weapon*, Sanger warns that the current cyber arms race is running without the same level of public debate or oversight of the Cold War nuclear age, where mutually assured destruction kept weapons locked away. 'Everything that worked in the nuclear age won't work for cyber,' he says now. 'Deterrence won't hold.'

The problem is that, in regular warfare, to deter an attack you must either be prepared to retaliate with a worse blow or make your attacker believe their assault was pointless, as your defences are too strong.

Neither is happening in cyberspace. Not only is cyber security weak across the board but nations are reluctant to strike back for fear of tipping cyber conflict closer to real war. They are also, despite the urging of experts, often unwilling to name and shame nations behind attacks.

'Imagine if we got it wrong and [blamed] the wrong country,' says Coyle.

In the shadows of cyberspace, states do not attack with national flags raised. To cover their tracks, they might even outsource hacks to criminals or cowboy civilians. Or an attack could be staged to look like ransomware (where criminals encrypt a computer's data then demand money to unlock it), when really destruction, not cash, is the goal.

Still, Austin insists governments everywhere are getting very good at attributing attacks, especially those sophisticated enough to be considered state-sponsored. 'It's mostly politics [and] fear of exposing sensitive intelligence sources or methods of our own that stops nations [pointing fingers].'

After all, countries under siege are usually themselves launching attacks. 'If Australia, the US, the UK go too far down the path of calling out every attack, China and Russia might start doing the same,' Austin says. 'So far, they only call out what they regard as attacks beyond the pale, such as Sony and Ukraine [NotPetya].'

In 2020, it was revealed that popular software had been used to infiltrate US government departments and companies around the world. Soon after taking office, President Joe Biden expelled Russian diplomats and sanctioned individuals and companies linked to the hack, known as the SolarWinds breach.

The world may not quite be in another Cold War but everyone agrees cyberspace will figure more prominently in conflict to come. Australia's first ambassador for cyber affairs, Dr Toby Feakin, told an international forum hosted by the ANU in 2021 that cyber had become central to foreign affairs in a 'way we never could have imagined'. Cyber capabilities and technology such as AI will 'fundamentally shape and shift the power dynamics of the twenty-first century', he said.

Sanger and others argue that the world now needs a digital Geneva Convention to rein in this Wild West – keeping civilian targets such as hospitals and power grids off-limits in a kind of 'cyber no-fly zone'. Austin says existing international law covers cyberspace in a sense but he agrees there are still critical questions to answer about how it can be applied. 'So, you can't bomb a hospital but you could disable its computer systems so people will die. For most people, that should break [rules of war] too.' Of course, as Sanger notes, while the big powers have agreed to some non-binding negotiations around cyber warfare, it may not be in their interest to muzzle their own capabilities.

'Another treaty would be just another piece of paper,' Uren says. 'We've had pretty successful prohibition of nuclear weapons because everyone is terrified of the consequences of using them, for good reason. The problem is people are not deadly terrified of the consequences of cyber. We either have to get better at defending ourselves or make the consequences worse for attackers.' Coyle agrees that getting the worst offenders to come to the table would be almost impossible, given they already refuse to admit to hacks. 'But if we could do it, it would be a wonderful thing.'

So, is the threat of cyberwar looming larger today than when Stuxnet was unleashed?

Austin says the attacks are certainly getting more vicious, and the hackers more resourced. But he thinks countries of all stripes will remain wary of putting the tens of trillions of dollars in the world's online banking system at risk with all-out cyberwar. 'Of course, it doesn't mean they can't navigate around it.'

What worries Coyle most is what she can't see coming. 'What's out there that we're not tracking? Has something been laid already? That's why we always have a presence in cyberspace, we're always war-gaming, so we're ready . . . But I'm an optimist, no one really wants to go to war.'

Uren, too, is hopeful cyber attacks will stay below the red line, even as he warns of escalating vulnerability in an increasingly connected world. 'On the whole, technology has made our lives better. There hasn't been some existential hit to our society, there hasn't been a catastrophe. At least, not yet.'

31

HOW COULD DUGONG FOOD HELP HUMANS SURVIVE?

Shy dugongs, green turtles and ancient, rocky microbes live among Australia's seagrasses and mangroves. Now politicians are interested in these ocean meadows. Why?

Emma Young

At Australia's westernmost point lies the Coral Coast, a land of strange extremities. Tourists swim with humpbacks and manta rays and whale sharks. They hand-feed dolphins that leap and play, obligingly sticking out scarred and blemished dorsal fins that let scientists identify them as individuals, name them and track their social behaviours. But nearby, the dugong's nose breaks the water only to breathe, and then it sinks back down, out of sight.

The sight of dugongs and their cousins, the manatees, once led sailors to believe they had seen mermaids, hence the name of the order they belong to: sirenia. But although dugongs are legendary, they are shy and cryptic creatures. Despite their imposing size, they don't emit growls or rumbles but chirp and squeak like tiny birds. They are slow to escape the paths of boats and liable to get injured. Rather than play, they prefer to lumber along the seafloor alone or in small groups in a finicky search for a favourite type of seagrass, a quest made harder by the fact they have terrible eyesight. They are perhaps at their most animated when they find the grass and use their teeth to rip it out, roots and all.

It is perhaps unsurprising that such a ponderous, vulnerable animal is globally declining, threatened with extinction as its numbers fall in the coastal waters of the Indian and Pacific oceans. Yet their ill-suitedness to this modern world has an exception. Western Australia's Shark Bay, on the Coral Coast stretching from Cervantes to Ningaloo, is the dugong's stronghold, boasting the largest stable population in the world. In this World Heritage Site, where red desert meets blue ocean, numerous other endangered species, and all-but extinct scientific curiosities, find refuge.

Of the world's 50 marine World Heritage Sites, Shark Bay is one of only four that meets every natural criterion for listing. And Shark Bay, Ningaloo and its neighbour Exmouth Gulf, along with the Great Barrier Reef, are shaping up as some of the world's most significant hotspots for blue carbon, one of the buzzwords

Prime Minister Scott Morrison has used in his response to calls for climate action.

But what actually is blue carbon, and can it save us?

What is blue carbon?

It's now common knowledge that trees store carbon that would otherwise pollute our atmosphere. What is less well known is that three types of marine vegetation found where land and sea meet – seagrass beds in shallow waters, mangrove forests in between, and tidal (salt) marshes covered in saltbush – absorb and store carbon 30 times faster than forest.

As with trees, the marine vegetation's roots and leaves store carbon, but the difference is the marine systems store most of the carbon in their soil (between 98 and 99 per cent for tidal marsh and seagrass, 60 per cent for mangroves). And unlike land forests, that soil is constantly building. It accumulates particles from the ocean, and sediment from run-off after storms on land. Seagrass also provides a home for countless shell-dwellers such as clams and snails, and when they die and decompose they form carbonate sands that further build the soils.

These processes also work to furnish beaches, elevate the shoreline and sea floor, and thus not only store carbon but physically counteract sea level rise, says Oscar Serrano, Edith Cowan University PhD research fellow and one of Australia's most eminent blue carbon experts. Serrano was one of the lead authors of a report published in 2021 by UNESCO, which for the first time assessed the vegetation fields in its 50 marine World Heritage Sites. Significant hotspots included the Everglades National Park in Florida, the West Norwegian fjords and the lagoons of New Caledonia, but none were as impressive or diverse as Australia's.

More than half the World Heritage Site blue carbon ecosystems in terms of size are in Ningaloo, Shark Bay and the Great Barrier Reef alone. All three contain tidal marshes, seagrasses and mangroves: Shark Bay and Ningaloo have significant extents of all three but particularly seagrass; Shark Bay has the world's biggest seagrass meadow; the Great Barrier Reef has vast swathes of all three. The report also quantifies the carbon locked away in these fields and shows they contain 40 per cent of the total blue carbon held in World Heritage marine areas.

Other blue carbon hotspots in Australia, Serrano says, are the mangrove forests of the Carpentaria and Exmouth gulfs; the seagrass meadows of Spencer and Vincent gulfs of South Australia; and the mangroves and tidal marshes of the whole tropical and subtropical region, encompassing Western Australia's north-east, the Northern Territory and Queensland. Together, Australian blue carbon ecosystems sequester 20 million tonnes of carbon dioxide a year. That's like taking four million cars off the road.

But globally, this blue carbon is being lost faster than tropical rainforests. Damage from coastal and marine development, severe weather and the effects of climate change are causing three million tonnes of blue carbon to be released back to the atmosphere each year.

> *Globally, this blue carbon is being lost
> faster than tropical rainforests.*

An extreme example was a 2011 seagrass die-off in Shark Bay caused by a marine heatwave, releasing up to 9 million tonnes of carbon dioxide and causing 'black water' in the region's famous tourist destination Monkey Mia. Another was the 2016 dieback deaths of about 7400 hectares of mangrove forest in the Gulf of Carpentaria, caused by extreme temperatures, drought and lowered sea levels.

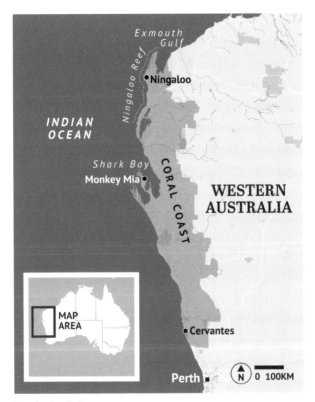

Map by Jamie Brown

When did blue carbon become a household phrase?

In April 2021, the prime minister called Australia's oceans 'part of the lifeblood of our economy' as he announced $100 million for ocean health including a $30 million blue carbon fund. The commitments, made as Morrison addressed 40 world leaders at US President Joe Biden's climate summit, reflect how blue carbon is being entered into the country's climate action political phrasebook.

As international pressure grows for Australia to set ambitious emissions reduction goals like its major trading partners the European Union, United Kingdom, United States and Japan, Australia has resisted – but proffered practical measures, including blue carbon.

Green carbon is already part of the machinery of carbon offsets and credits used to counteract pollution. The Australian government has a dedicated Emissions Reduction Fund to pay companies carbon credits to offset their pollution through planting trees. It's now working on an accounting method, planned for completion in 2022, so polluters can claim credits through protecting or restoring blue carbon ecosystems.

Serrano, who is advising the government on how blue carbon projects can be banked against emissions reduction targets, says in future any big polluter will need to invest in both blue and green offsets to reach their stated carbon abatement goals. He says the capacity of blue carbon per square metre is vastly greater than green. 'Soils underneath the forest also have carbon but once the forest is mature the soil can't accumulate more,' he says. 'Most of the carbon is in . . . the trees themselves. And they don't live forever. They live 50, 100 years. If you lock away carbon you want it to be locked away for centuries.'

Projects are contemplated only for tidal marsh and mangrove forest. Seagrass is harder to account for as it can't be mapped with aerial imagery, requiring instead boats and human labour, which are expensive. But Australia is one of the first countries in the world to get even this far. 'My feeling is the Australian government is wanting to take the lead worldwide on this and they are starting with this, but soon other methods will follow and hopefully they will include seagrass,' Serrano says.

Blue carbon may be an effective diplomatic tool for Australia which has been under pressure to act on climate action from Pacific leaders; the government's $30 million for blue carbon includes $10 million to support three restoration projects in developing countries, most likely in the Pacific.

Some Australian states are taking action of their own, too. Queensland is investing in blue carbon through its Land Restoration Fund, South Australia is developing a blue carbon strategy. In Western Australia, climate action advocacy organisation Clean State has submitted to the government a $12 million blue carbon plan they say would create 270 jobs.

Can blue carbon help tackle climate change?

Conserving, protecting and rehabilitating marine habitats is a worthy endeavour, and blue carbon an interesting and legitimate scientific field, says Monash University biogeochemist Perran Cook. Unquestionably, destruction of land and marine vegetation releases carbon dioxide and further imbalances the climate budget. But the professor does not view blue carbon – or green – as a realistic way to combat fossil fuel emissions. 'When it's used politically in such an important policy area, it's hanging a hat on something very uncertain,' he says.

The first question is, can blue carbon restoration and protection balance our carbon budget? The 20 million tonnes of CO_2 that Australian blue carbon systems lock away a year is equivalent to less than 6 per cent of the emissions from fossil-fuel burning. Australia's ten biggest polluters, all fossil fuel companies, together emit more than 150 million tonnes a year alone. 'The amount of CO_2 we are emitting, the rate at which you'd have to rehabilitate [marine] areas would be enormous,' Cook says.

The second question is, is it practical? Cook says, as with any marine engineering, restoration of blue carbon ecosystems is relatively expensive. And, land or sea, not all offsets are created equal. Forests might be a high-quality permanent offset in one area, such as around Cairns with its high rainfall, but not in eastern Australia where climate change might one day turn them into savannah. He suspects a back-of-the-envelope calculation might show Australia would 'really struggle' to repurpose enough agricultural land for tree planting on the scale required. 'We have turned a lot of forest into agricultural land,' he says. 'We have burnt a huge amount of fossil fuels. By replanting it all . . . you can't offset a lot of the burning of fuels, just the carbon released from cutting down those trees in the first place.'

The third issue is stability. Cook questions the idea of blue carbon having a longer lifespan in times when climate change, marine heatwaves and rising sea levels are destabilising oceans. He cites those devastating effects of marine heatwaves on the Gulf of Carpentaria mangrove forests in 2016 and Shark Bay seagrass meadows in 2011. 'You'd have to be able to say it has a long-term permanency to the atmosphere on a timescale of up to hundreds of thousands of years,' he says. 'The permanency of carbon in those supposed offsets is far from guaranteed.'

Serrano agrees Australia cannot rely on blue carbon alone. 'Even if we restore all blue carbon ecosystems in Australia these would only contribute a small portion of the necessary abatement,' he says. 'Blue carbon needs to be part of the solution. But the solution is cutting emissions.'

What else is at stake?
Dugongs are not the only unique life forms along the Coral Coast at risk from climate change. Other endangered species rely on seagrass, such as green turtles. Dependent on the turtles are the tiger sharks that give Shark Bay its name.

Also in Shark Bay, the world's only significant assembly of live stromatolites survive – reefs of layered rock-and-microbe domes, once the planet's dominant life form, before the creation of oxygen rendered them extinct virtually everywhere but this sheltered, shallow and hyper-saline site. NASA has studied them as part of a search for life on other planets. And like the water surface that gives little clue to the 10,000 dugongs cruising beneath, the single tourist boardwalk gives little perspective on the stromatolite reefs' 135-kilometre expanse, on which only the odd researcher is permitted. In Monkey Mia, another area of Shark Bay, another oddity: a tiny cultivated pod of dolphins visits the beach daily to interact with humans, achieving world fame among both tourists and scientists.

At the other end of the Coral Coast, 400 kilometres north, Exmouth Gulf beside Ningaloo Reef is the only known site worldwide where green sawfish give birth to pups. A five-metre-long, shark-like creature whose toothed snout, as long as its body, resembles a chainsaw blade, it looks fearsome but is one of the planet's most endangered species.

Unseen along this stretch of coast are 'stygofauna' found nowhere else but Western Australia: the blind cave eel and blind gudgeon fish live deep underground in a complex terrain of limestone caves, rivers and lakes.

The peninsula on which Exmouth sits has fossilised coral reefs that date back to the last Ice Age, and whose very form has contributed to this staggering biodiversity, by sheltering the Gulf from the rest of the ocean for millennia. And this is just what is already known. The only region-wide survey of Exmouth Gulf done in recent years, by Oceanwise Australia, found science had barely scratched the surface; despite this, the Gulf likely would already qualify for World Heritage status like its neighbour Ningaloo.

Rising sea levels, marine heatwaves and cyclones threaten all the curiosities found in these blue carbon hotspots, and those across the rest of Australia.

Meanwhile, happily oblivious in Moreton Bay, Queensland, one of Australia's smaller populations of dugongs is even more idiosyncratic than their western counterparts. They form vast herds to raze their preferred seagrass variety annually, to encourage regular, tender and delicious growth: a process so uncannily like farming that experts call it 'cultivation' grazing. Yet they cause no harm.

Unless we, too, can find a positive way to work with nature, the interconnectedness of all these systems suggests humanity stands to lose more than just safety and comfort to climate change: we stand to lose wonder, too. Preserving blue carbon ecosystems might not be the complete answer. But it could be a start.

*

With thanks to Dr Amanda Hodgson, aquatic megafauna research fellow at Murdoch University, for her insights into the life of a dugong.

Author biographies

Sophie Aubrey reports on everything from personal health science and wellness quackery to social trends and digital culture as deputy lifestyle editor for *The Age* and *The Sydney Morning Herald*. She has worked at the *Herald Sun, mX* and News Corp's national network as well as for digital publication *Mamamia*. Her research into ageing has convinced her to never give up exercising – or at least, she'll try not to – and that ageing is a gift we should embrace and plan for early on.

Eryk Bagshaw is based in Singapore as the North Asia correspondent for *The Sydney Morning Herald* and *The Age*. Between covering geopolitics, territorial disputes and COVID-19, he plans on eating his way through the region. A former federal political reporter and economics correspondent in Canberra, he was the Wallace Brown Young Press Gallery Journalist of the Year in 2019. He was also the winner of an Our Watch Walkley Award in 2017 and the Walkley Young Australian Print Journalist of the Year in 2016.

Elizabeth Farrelly is a Sydney author and columnist who is trained in architecture and philosophy and holds a PhD in urbanism from the University of Sydney. She is a former associate professor (practice) at UNSW Graduate School of Urbanism and a former City of Sydney alderman and councillor. A weekly columnist for *The Sydney Morning Herald*, she has published several books including *Blubberland: The Dangers of Happiness* (2007) and *Killing Sydney: The Fight for a City's Soul* (2021). She is building an off-grid dwelling in rural New South Wales whose spirit-animal is a barn and whose plan, naturally, is part open, part closed.

Mike Foley can be found kayaking up freshwater streams in the east coast hinterland looking for Australian bass, hiking into northern rainforest to fish for elusive jungle perch, or casting oversized lures for mighty Murray cod in the Murray–Darling Basin in his spare time. As the environment and energy reporter for *The Sydney Morning Herald* and *The Age*, at federal parliament in Canberra, he has a particular interest in all things policy related to rivers, oceans and water. Mike has written about the natural environment for the past decade. He was previously a rural reporter for *The Land* newspaper.

Dionne Gain is the creative director at *The Sydney Morning Herald* and has been a leading illustrator for the *Herald* for more than fifteen years, her work covering subjects from political commentary to lifestyle. She illustrated the first explainer anthology, *What's It Like to Be Chased by a Cassowary?* in 2020. She was highly commended in the Warringah Art Prize when she was twelve and has a penchant for jumpsuits.

Sherryn Groch is the national explainer reporter for *The Age* and *The Sydney Morning Herald*, digging into everything from the dark history of political poisonings and the secret lives of sharks to the surprisingly poignant question of time travel. She previously covered crime, education and social affairs for *The Canberra Times*, for which she won a Walkley Young Australian Journalist of the Year Award in 2020. She really likes sharks.

Patrick Hatch likes to fly abroad when he can and hopes that will be possible without carbon emissions – or *flygskam* (flight shame) – in his lifetime. His special focus is on aviation and transport in his role as business reporter for *The Age* and *The Sydney Morning Herald*. He has worked at *The Age* since 2013 as a breaking and general news reporter and digital editor. Patrick joined the national business team in 2015 where he has also written about the retail, healthcare and gambling industries.

Melanie Kembrey writes about literature, art and popular culture as culture deputy editor at *The Sydney Morning Herald* and *The Age*. She was previously deputy editor of the *Herald*'s Saturday arts section Spectrum, and has covered courts and crime. She can usually be found beneath a pile of books or wandering around an art gallery.

Sarah Keoghan is a sports reporter for *The Sydney Morning Herald*. She loves playing soccer and watching rugby league but is yet to try her hand at breakdancing.

Felicity Lewis is the national explainer editor for *The Age* and *The Sydney Morning Herald*, and edited the first explainer anthology *What's It Like to Be Chased by a Cassowary?* She has reported on everything from crime to interior design, geopolitics to lobster rolls in more than 25 years as a journalist, working in diverse roles on mastheads from the *Herald Sun* to *The Independent* to *theage(melbourne)magazine,* and has won awards including a Walkley.

Phil Lutton writes about rugby league and rugby union, cricket, golf and tennis as a senior sports reporter for *The Sydney Morning Herald* and has a special affinity for the Olympic Games, covering his first in Athens in 2004. He started writing about swimming at the London Olympics in 2012 and has gone on to break some big stories in aquatics, one of which sparked Mack Horton's famous showdown with Chinese rival Sun Yang in Rio in 2016. He lives in Brisbane and spends half his summers in pools or at the beach.

Garry Maddox is a long-time specialist in film who has been covering the Academy Awards – with a strong record of predicting the winners of best picture – for more than two decades. A senior writer for *The Sydney Morning Herald* and a regular host of film Q&A sessions, he has profiled the likes of George Miller, Baz Luhrmann, Nicole Kidman, Cate Blanchett, Ivan Sen, Warwick Thornton, Mel Gibson and Alan Rickman. He has also been policy manager for the Film Finance Corporation, author of a series of film- and television-industry reports, a playwright and board member of Sydney Film Festival.

Konrad Marshall is a senior writer with *Good Weekend* magazine, where his long-form articles often focus on sport. His March 2019 cover story on concussion won a Melbourne Press Club Quill Award for sports writing, and was also part of a portfolio of work that saw him named Harry Gordon Australian Sports Journalist of the Year. A football tragic, Marshall would dearly love to have played in the ruck for Richmond, but recognised early in life that tall men who cannot mark the ball have precious little utility. Admirably, he chanelled those frustrations into three bestselling books on the Tigers' 2017, 2019 and 2020 premiership campaigns.

James Massola has worked in Parliament House covering federal politics since 2008, except when he was South-East Asia correspondent for *The Age* and *The Sydney Morning Herald* from 2018 to 2020. He won the NRMA Kennedy Award for Outstanding Foreign Correspondent in 2019 and wrote *The Great Cave Rescue*, a book about the rescue of a Thai boys' soccer team from Tham Luang cave in 2018. He is now the *Sunday Age* political correspondent. When he's not knocking on politicians' doors or bailing them up in the Senate courtyard, his pastimes are gardening, reading and building Lego with his three young kids.

Nick Miller curates daily arts coverage and reports on the trends, performers and hot tickets in Melbourne's cultural scene as arts editor for *The Age*. Prior to 2020, he was Europe correspondent for *The Age* and *The Sydney Morning Herald*, where he found time to visit movie sets and chat to actors, dancers and musicians in between the terror attacks and geopolitics. He was, very briefly, a long time ago, a professional actor and would-be playwright, though his only play staged in Sydney was panned by... *The Sydney Morning Herald.*

Maher Mughrabi was born in Scotland, the son of a Palestinian father and a Scottish mother. He has written and edited for newspapers for over 25 years, starting in Dubai before working in Aberdeen, Edinburgh, Glasgow and London. He migrated to Australia in 2003 and joined *The Age*, moving to the foreign desk in 2007 and becoming the *Age* and *Herald's* foreign editor from 2014 to 2017. Since September 2017 he has been features editor of *The Age*.

Julia Naughton usually prefers to tell other people's stories, but dissecting her own for her explainer proved insightful, confirming her scepticism about the Disney narrative of love – attraction is fleeting, but love is something that deepens over time. Naughton leads coverage of fashion, beauty, relationships and personal health as the lifestyle editor for *The Sydney Morning Herald* and *The Age*. She was previously the managing editor of nine.com.au where she wrote and produced the royal podcast *The Windsors*, and has worked on titles such as HuffPost Australia and *Cosmopolitan*.

Iain Payten is deputy editor of sport for *The Sydney Morning Herald*. He previously worked for *The Australian*, the *Daily Telegraph* and the *Daily Mail* in England. He has covered two Olympics and rates Chloe Esposito's gold medal in the modern pentathlon in Rio in 2016 as the best Olympic performance by an Australian he has covered as a journalist – closely followed by the Australia women's rugby sevens team winning gold. Payten possesses no ability to breakdance and should be discouraged from doing so if observed trying.

Dominic Powell is a recovering bitcoin fanatic who definitely doesn't regret selling all his cryptocurrency at the end of 2018. A business journalist with *The Age* and *The Sydney Morning Herald,* he has a particular interest in retail companies. When not pondering his lost millions, he enjoys cooking, film photography and playing the collectible card game Magic: The Gathering.

Karl Quinn traded his Brummie accent (that's Birmingham, orright) for an Aussie one as the five-year-old child of 10-pound Poms and has been fascinated by accents ever since. Sadly resigned to the fact he probably shouldn't do them around the office any more, he nonetheless retains a keen ear for people's endlessly varying speech patterns, and an interest in the forces that shape them. He is senior culture writer at *The Age* and *The Sydney Morning Herald*.

Samantha Selinger-Morris was born in Toronto and moved to Sydney 23 years ago where she is still learning to cope with the harsh sun. She spoke with a range of experts for 'What does the sun do to your skin?' and it's clear what's needed: extreme vigilance. Selinger-Morris reported for *The Globe and Mail* in Canada and the ABC and SBS in Australia before joining *The Sydney Morning Herald* and *The Age*. As for the so-called psychedelics renaissance she also explains in this book, she was surprised that drugs best known for making people believe they can fly are now showing great promise in the treatment of conditions from alcoholism to PTSD.

John Silvester is a Walkley Award–winning crime writer and Naked City columnist for *The Age*. An earlier version of 'How do you stop crimes of fixation?' was first published in his column. He has written and edited more than 30 crime books and is the co-author of the bestselling *Underbelly* series that was the basis of the hit Australian TV series. Silvester is a regular guest on 3AW with his Sly of the Underworld segment.

Jewel Topsfield covers everything from mental health and family violence to menopause being the last workplace taboo as the social affairs editor for *The Age*. Her previous roles include education editor, Melbourne editor, and Indonesia correspondent from 2015 to 2018, when she won a Walkley Award for International Journalism and the Lowy Institute Media Award. Researching 'What is it like to live with dementia?' taught her, in the words of one of the carers she interviewed, to 'live in the present a bit more and look for moments of creating joy'.

Matt Wade was based in India as a foreign correspondent for *The Sydney Morning Herald* and *The Age* between 2007 and 2011. He met the Dalai Lama at a press conference in north India while covering anti-China protests by Tibetans in 2008. Wade, a senior writer who covers economics, politics and demography, has twice won the ACFID Media Award for his coverage of international development and humanitarian issues in Africa and the Pacific. He previously worked in Canberra as the *Herald*'s economics correspondent.

Emma Young has visited Western Australia's Coral Coast and reported extensively on its environmental values and the challenges faced by the region. A reporter for *The Age* and *The Sydney Morning Herald*'s sister publication *WAtoday*, she has won eight WA Media Awards, including the 2018 Matt Price Award for best columnist and awards for environmental reporting. She also writes fiction. Her debut novel, *The Last Bookshop*, was shortlisted for the inaugural Fogarty Literary Award in 2019 and was published by Fremantle Press in 2021.